ANIMAL ECOLOGY

ANIMAL ECOLOGY

BY

CHARLES ELTON

WITH NEW INTRODUCTORY MATERIAL BY

MATHEW A. LEIBOLD

AND

J. TIMOTHY WOOTTON

THE UNIVERSITY OF CHIGAGO PRESS

CHICAGO & LONDON

TO MY BROTHER

GEOFFREY ELTON

CHARLES ELTON established and led Oxford University's influential Bureau of Animal Population. He was the author of a number of books, among them *The Ecology of Invasions by Animals and Plants*, recently reprinted by the University of Chicago Press.

MATHEW A. LEIBOLD is associate professor in the Department of Ecology and Evolution and chair of the Committee on Evolutionary Biology at the University of Chicago.

J. TIMOTHY WOOTTON is associate professor in the Department of Ecology and Evolution and on the Committee on Evolutionary Biology at the University of Chicago.

Published by arrangement with Kluwer Academic Publishers, B. V.
The University of Chicago Press, Chicago 60637
First published in 1927
©1966, Sidgwick & Jackson, Chapman & Hall,
Kluwer Academic Publishers, B. V.
New Introductory Material ©2001 by The University of Chicago

The University of Chicago Press edition 2001
Printed in the United States of America

09 08 07 06 05 04 03 02 01 00   1 2 3 4 5

ISBN 0-226-20639-4 (paperback)

Library of Congress Cataloging-in-Publication Data
Elton, Charles S. (Charles Sutherland), 1900–
Animal ecology / Charles Elton ; with new introductory material by
Mathew A. Leibold and J. Timothy Wootton.
p.   cm.
Originally published: 1927.
Includes bibliographical references and index (p.   )
ISBN 0-226-20639-4 (pbk. : alk. paper)
1. Animal ecology.   I. Title.
QH541.E398   2001
591.7 21; aa05 12–20—dcoo   00-069087

⊗The paper used in this publication meets the minimum requirements of the American National Standard for Information Sciences—Permanence of Paper for Printed Library Materials, ANSI Z39.48-1992.

CONTENTS

LIST OF PLATES AND
DIAGRAMS IN THE TEXT

AUTHOR'S PREFACE

ECOLOGICAL methods can be applied to many different branches of animal biology. For instance, they may be employed in the study of evolution and adaptation. Most of the work done so far under the name of ecology has been concerned with this side of the subject, and has been summed up to some extent by Borradaile (*The Animal and its Environment*) and Hesse (*Tiergeographie auf Ökologische Grundlage*). Or they may be applied to the study of the normal and abnormal life of cells in the bodies of animals, as has been so strikingly done by Morley Roberts (*Malignancy and Evolution*) ; or again they may be used in the study of man, when they form the sciences of sociology and economics (as exemplified by Carr-Saunders in *The Population Problem*), or of the relations between the two sexes, as illustrated by the work of Eliot Howard and J. S. Huxley on bird-habits. The present book is chiefly concerned with what may be called the sociology and economics of animals, rather than with the structural and other adaptations possessed by them. The latter are the final result of a number of processes in the lives of animals, summed up over thousands or millions of years, and in order to understand the meaning and origin of these structures, etc., we must study the processes and not only their integrated results. These latter are adequately treated in the two works mentioned above, and although of great intellectual interest and value, a knowledge of them throws curiously little light on the sort of problems which are encountered in field studies of living animals. I have laid a good deal of emphasis on the practical bearings of many of the ideas mentioned in this book, partly because many of the best observations have been made by people working on economic problems (most of whom, it may be noted, were not trained as professional zoologists), and partly because the principles of animal ecology are seldom if ever mentioned in zoological courses in the universities, in spite of the fact that it is just such knowledge which is required by any one who is brought up against practical problems in the field, after he leaves the university. Ecology is a branch of zoology which is perhaps more able to offer immediate practical help to mankind than any of the others, and in the present rather parlous state of civilisation it would seem particularly important to include it in the training of young zoologists. Throughout this book I have used analogies between human and animal communities. These are simply intended as analogies and nothing more, but may also help to drive home the fact that animal interrelations, which after all form the more purely biological side of ecology, are very complicated, but at the same time subject to definite economic laws.

Finally, I wish to point out that no attempt has been made to provide a handbook containing references to all the ecological work that has ever been done, since such a work would be both exhausting to read, and useless when it had been read. I have simply taken the various principles and ideas and illustrated them by one or two examples.

I am indebted to Mr. O. W. Richards for a great deal of help and criticism. Many of the ideas in this book have been discussed with him, and gained correspondingly in value, and in particular his extensive knowledge of insects has been invaluable in suggesting examples to illustrate various points. The list of works dealing with British insects was compiled with his aid.

I am also indebted to Professor J. S. Huxley for much helpful advice, and I have to thank Dr. T. G. Longstaff, Mr. J. D. Brown, Dr. K. S. Sandford, and Captain C. R. Robbins, for allowing me to use photographs, all taken in rather out-of-the-way parts of the world (Spitsbergen, Egyptian desert, and Burma, respectively).

CHARLES ELTON.

University Museum,
Oxford.

EDITOR'S INTRODUCTION

WHEN I decided to undertake the editing of this series of volumes, I had a perfectly definite idea in mind. Biological science has been of late years growing and expanding at a prodigious rate. As a result, teachers of zoology and also of botany—but I shall confine myself to Animal Biology—have had to face the gravest difficulties in regard to their curriculum. The first difficulty is a purely quantitative one : now that the subject has invaded so many new fields, how to stuff this tenfold bulk of knowledge into the brains of students in the same time as before. The second difficulty concerns the relative value of the different biological disciplines. Shall Comparative Morphology continue in the future to dominate the undergraduate's learning period, with Genetics, Cytology, *Entwicklungsmechanik*, Animal Behaviour, Systematics, Distribution, Ecology, Histology, Comparative Physiology, and Evolution tacked or thrown on here and there like valances or frills or antimacassars ? or can it and should it renounce its pretensions and become one of a society of equals ?

It is to my mind more important to attempt an answer to this second question first. I do *not* believe that comparative morphology has the right to demand the lion's share of the students' time and energy. That it at present obtains that lion's share is due almost entirely to historical reasons—to the fact that zoological departments grew up while comparative anatomy and morphology were the most fruitful and the most interesting lines of attack in zoology.

Those who uphold the present system tell us that morph-ology is the foundation of zoology, the backbone of the subject, and that it is impossible or unprofitable to embark on subjects

like comparative physiology or developmental physiology or
ecology until the student has gained a general knowledge of
the main types of the animal kingdom. I am quite prepared
to admit that morphology is the backbone of zoology. On
the other hand, I am not at all sure that it is the foundation of
our science; I should be more disposed to confer that title
upon physics and chemistry. However, all such doubts
apart, the fact of being either a foundation or a backbone most
emphatically does not call for the size-privileges at present
accorded to morphology. We do not live in the foundations
of our houses, nor are they larger than the superstructure.
And as for backbones, it should not need more than a very
elementary acquaintance with natural history to realise that
an animal whose backbone weighed more than its muscles,
nervous system and viscera combined, would be biologically
very inefficient.

As to the claim that other subjects can only be tackled
after a morphological survey of the animal kingdom, this must
be taken *cum grano*. It is in one sense obviously true, but in
another completely false. It is false if the knowledge of
morphology assumed is that detailed and intensive knowledge
which is usually required for a zoological degree. It is true
if we mean that a general survey of the main types of structure
and development found among animals is a desirable pre-
requisite to many other branches of biology. But such a
survey can be given in a small fraction of the time now allotted
to the morphological discipline; what is more, if thus given,
the wood will not be obscured by the trees, which is unfortunately too often the case (*experto crede!*) when the intensive
and detailed system is practised. There are only about twelve
phyla in the animal kingdom; while the total number of groups,
whether sub-phylum, class, sub-class or order, of which the
budding zoologist need know the bare existence before he
embarks on general biology, is certainly less than a hundred,
and the characteristics of at least half of these he need only be
acquainted with in the most superficial way, provided that he is
well instructed in the ground-work of the phyla and sub-phyla.
Another claim which I have often heard made is that the

majority of the subjects of general biology can be most profitably treated in relation to a thorough general morphological survey of the animal kingdom. This, I must confess, appears to me pedagogically a most pernicious doctrine. It was all very well so long as the other subjects remained scrappy. But once they have penetrated deep enough to acquire definite principles of their own, the defects of the method are revealed, since it is almost impossible to teach two not very closely related sets of principles simultaneously. This was soon recognised for subjects with a definite body of principles of their own, such as cytology or genetics; but the old point of view too often lingers in respect of, for instance, ecology, or systematic and faunistic studies.

The remedy, in my opinion, is to drop the whole notion of having a single main course around which, like his paraphernalia around the White Knight, the remainder of the subjects are hung. There should be a series of courses of approximately equal "value," each covering one of the main fields of biological inquiry, each stressing a different set of principles, so that the student will at the close have seen his science from the greatest possible number of different angles. As a preliminary programme, I should suggest about ten such courses. For instance : (1) vertebrate morphology, stressing the principles of comparative anatomy; (2) vertebrate embryology, stressing the principles of development, including organogeny and histogenesis; (3) the invertebrates and lower chordates, stressing both comparative anatomy and embryology, so as to bring out the divergencies and various grades of animal life; (4) cytology and histology; (5) genetics; (6) developmental physiology, including the effects of function upon structure, regeneration, dedifferentiation, tumour-formation, etc., as well as what is usually called experimental embryology; (7) faunistic zoology and ecology, bringing out the types of environment, and of adaptations to various environ-ments, as well as the "animal sociology and economics" covered by ecology in the narrower sense; (8) comparative physiology; (9) animal behaviour; and (10) evolution, including some treatment of the principles of systematics.

I do not wish to imply that each of these courses should require the same number of lectures, still less the same amount of practical work. But I do claim that each of them is in a certain real sense of equal importance, since each of them, if properly taught, can impart its own characteristic point of view ; and I claim that each of these points of view is of equal importance to any one claiming the title of biologist.

It would be perfectly possible to add to the list : Economic Zoology and Historical Zoology at once occur to the mind. It would be equally possible to arrange for a divergence in specialisation in a student's last year, some choosing the more physico-chemical, some the more biological side of the subject. But such details must depend on the experience gained as teaching adapts itself to the growth of the subject, and also upon local conditions.

Thus what I had in mind in arranging for this series, was to make an attempt to cover these separate fields, each field being handled on approximately the same scale. Some modern developments have been so well treated in recent years that I have not thought it worth while to enter into useless competition with the admirable existing works. That is eminently the case with genetics, in which the volumes by Crew, Morgan, Jones, and others already carry out what was in my mind.

Other fields have also been covered, but covered too well. Parker and Haswell, Adam Sedgwick, MacBride's *Invertebrate Embryology* and Graham Kerr's *Vertebrate Embryology*, are books of reference. To ask undergraduate students to read through such works is merely to give them mental indigestion, though they may obviously be used with great profit in conjunction with lecture-notes and short text-books.

T. H. Huxley, that great scholar and man of science whose name I am proud to bear, wrote one text-book of Vertebrate, another of Invertebrate Anatomy. It is text-books of that size and scope at which this series aims, in which the detail shall be used to illustrate the principles, but no deadening attempt made at a completeness which in any case must remain unattainable.

Finally, there remain subjects which are of such recent growth that their principles have never yet been treated in a comprehensive way. Such, for instance, are developmental and comparative physiology, animal behaviour, and ecology. From the point of view of the rapid growth and expansion of general biology, it is these subjects which it is at the present moment most important to summarise in brief text-books, since otherwise the multifarious knowledge which we have already attained regarding them remains locked up in scattered papers, the property of the specialist alone.

The present volume deals with a much misunderstood and often underrated subject. If we leave out Hesse's *Tiergeographie auf ökologischen Grundlage*, which deals with faunas and major habitats and animal adaptations rather than with ecology *sensu stricto*, hardly any books dealing with the subject have been published since Shelford's fine pioneering work of 1913, *Animal Communities of Temperate America.*

The subject is also so new and so complex that it is only of recent years that principles have begun to emerge with any clearness. It has further suffered from taking over too wholeheartedly the concepts of plant ecology and applying them directly to animals instead of seeing whether the difference between animal and plant biology did not of necessity introduce a difference in the principles governing animal and plant ecology.

Mr. Elton, ever since I had the good fortune to have him as my pupil at Oxford, has been largely occupied with the problems of animal ecology and the quest for guiding principles in the subject. He has been fortunate in having field experience in the Arctic, where the ecological web of life is reduced to its simplest, and complexity of detail does not hide the broad outlines. He has also been fortunate in early becoming preoccupied with the subject of animal numbers ; or, I should rather say, he early showed characteristic acumen in seeing the fundamental importance of this problem. He is finally fortunate in having an original mind, one which refuses to go on looking at a subject in the traditional way just because it has always been looked at in that way. The result, it appeared to me as I read through his manuscript, is an illuminating and original book, the first in which the proper point of view of animal ecology has yet been explicitly stated. I will take but one example, and that from Mr. Elton's pet subject, the regulation of animal numbers.

Men of science do not escape the usual human weakness of regarding facts in a naïve and superficial way until some special stimulus to deeper analysis arises. I suppose that most professional biologists think of the relation of carnivores to herbivores, preyer to preyed-upon, almost wholly in the light of the familiar metaphor of *enemies*; and of the relation between the two as being in some real way like a battle. The ecologist, however, speedily arrives at the idea of an optimum density of numbers, which is the most advantageous for the animal species to possess. He then goes on to see by what means the actual density of population is regulated towards the optimum; and finds that in the great majority of cases the existence of enemies is a biological necessity to the species, which without them would commit suicide by eating out its food-supply. To have the right "enemies," though it can hardly be spoken of as an adaptation, is at least seen to be a biological advantage.

Ecology is destined to a great future. The more advanced governments of the world, among which, I am happy to say, our own is coming to be reckoned, are waking up to the fact that the future of plant and animal industry, especially in the tropics, depends upon a proper application of scientific knowledge. Tropical Research Stations, like those at Trinidad and Amani; special investigations, like that into the mineral salt requirements of cattle in equatorial Africa; schemes for promoting the free flow of experience and knowledge from problem to problem and from one part of the world to another, such as were outlined in the report of the Research Committee of the Imperial Conference—all these and more will be needed if man is to assert his predominance in those regions of the globe whose climate gives such an initial advantage to his cold-blooded rivals, the plant pest and, most of all, the insect.

To deal with these problems, a cry is going up for economic entomologists, mycologists, soil biologists, and the rest. *Ad hoc* training in these and similar subjects is being given at various centres, and special laboratories are being erected for research in the separate branches. Valuable results are being achieved : but the general biologist is tempted to ask whether in the quest for specific knowledge and specific remedies it is not being forgotten that behind all the detail there is to be sought a body of general principle, and that all these branches of study are in reality all no more and no less than Applied Ecology. The situation has many points of resemblance to that which obtained in medicine in the last half of the nineteenth century. Then, under the magic of the germ-theory and its spectacular triumphs, medical research on disease was largely concentrated upon the discovery of specific " germs ", and their eradication. But as work progressed, the limitations of the mode of attack were seen. Disease was envisaged more and more as a phenomenon of general biology, into whose causation the constitution and physiology of the patient and the effects of the environment entered as importantly as did the specific parasites.

So it will be with the control of wild life in the interest of man's food-supply and prosperity. The discovery of the tubercle bacillus has not led to the eradication of tuberculosis : indeed it looks much more likely that this will be effected through hygienic reform than through bacteriological knowledge. In precisely the same way it may often be found that an insect pest is damaging a crop ; yet that the only satisfactory way of growing a better crop is not to attempt the direct eradication of the insect, but to adopt improved methods of agriculture, or to breed resistant strains of the crop plant. In other words, a particular pest may be a symptom rather than a cause ; and consequently over-specialisation in special branches of applied biology may give a false optimism, and lead to waste of time and money through directing attention to the wrong point of attack.

The tropical entomologist or mycologist or weed-controller will only be fulfilling his functions properly if he is first and foremost an ecologist : and I look forward to the time when all the present *ad hoc* branches of applied biology will be unified in relation to laboratories of pure and applied ecology.

I will give but one example of the value of ecological knowledge and the ecological outlook in these matters. It is a familiar fact that serious plagues of mice, rats, and other rodents occur from time to time in various parts of the world, often causing a great deal of material damage. At the moment that I write these lines, the newspapers record a rodent plague in California so serious that all crops are in danger over a considerable section of the State. Readers of Mr. Elton's book will discover that these violent outbreaks are but special cases of a regular phenomenon of periodicity in numbers, which is perfectly normal for many of the smaller mammals. The animals, favoured by climatic conditions, embark on reproduc-tion above the mean, outrun the constable of their enemies, become extremely abundant, are attacked by an epidemic, and suddenly become reduced again to numbers far below the mean. When such a number-maximum is so accentuated as to become a plague, remedial measures are called for locally, and large sums of money may be spent. Eventually the normal epidemic breaks out and the plague abates. The organisers of the anti-rodent campaign claim the disappearance of the pest as a victory for their methods. In reality, however, it appears that this disappearance is always due to natural causes, namely, the outbreak of some epidemic ; and that the killing off of the animals by man has either had no effect upon the natural course of events, or has delayed the crisis with the inevitable effect of maintaining the plague for a longer period than would otherwise have been the case ! In the latter event, it would actually have been a better counter-measure to do nothing at all than to spend time and money in fruitless killing. If remedial measures are to be desired, they must be of some special sort. Either they must encourage the development of the epidemic, as by introducing infection among the wild population of the pest species ; or they must aim at reducing reproduction, as in the Rodier anti-rat campaign, where after trapping, only females are killed and all males liberated once more ; or they must be aimed at the general ecological status of the species, making it more difficult for it to live and reproduce, as has in another sphere been accomplished by drainage and cultivation with regard to the malarial mosquito.

I recommend Mr. Elton's book to biologists as a valuable and original contribution to pure science, and as a fresh foundation for applied zoology.

JULIAN S. HUXLEY.

February, 1927.

INTRODUCTION
Mathew A. Leibold and J. Timothy Wootton

Animal Ecology was last printed in 1968. Nevertheless this book, written in 1927 by a young man (Charles Elton was only 26 at the time) in a moment of passionate inspiration (Elton wrote the book in a period of only three months), has been among the most influential in the field of ecology. In part this is because it helped to formulate so many of the basic questions that ecologists consider "important" even to this day. As graduate students we were both encouraged to read Elton's book and we have both encouraged our own students to read it. Its influence also arises because, unlike most subsequent textbooks which largely review ecological facts and concepts in a coherent framework (Eugene Odum's tremendously influential 1953 *Fundamentals of Ecology* being a notable exception), Elton presents his own unique and highly synthetic view of how to approach the field. There persistently remains a place for such individualistic treatments of ecology, and we hope that the pressure to produce texts which are commercially palatable to the widest possible audience is not suppressing more synthetic, if less comprehensive, efforts in the field.

There are many great insights to be found in this book. Many of the basic lessons Elton tries to develop still apply today, including the importance of natural history, the use of multiple lines of inquiry, and the focus on the complex interactive nature of ecological systems. There is also great value in recognizing the historical foundations of our field, which is especially true when the issues and controversies that shaped Elton's thinking are described in such passionate and provocative terms herein.

Nevertheless, the field of ecology has undergone tremendous growth since 1927, and perhaps especially, in the last 15 years or so.

Today's emphasis on theoretical, statistical, and methodological issues are hardly to be found in Elton's book take on additional significance given work in the field done more recently. Remarkable also is the recent return of academic ecology to concerns with applied issues such as conservation and management strategies in ways that would probably resonate well with Elton.

Because we felt that Elton's *Animal Ecology* was such a vital part of the basic literature in ecology, we lobbied the University of Chicago Press to reprint it (plus we were tired of having to loan out our own copies or to borrow the book from the library!). However, because we also felt that much of what is most interesting in *Animal Ecology* requires some background in the field, we wanted to provide a guide to the book. We have done so in the form of short introductory essays, and we have tried to provide references to more recent studies that can provide links with some of the most active areas of current work. These essays reflect our own perspectives on the topics of each of the chapters. They strongly reflect our own interests, which we hope are not too narrow. No doubt, others would have probably provided very different perspectives that might have been at least as interesting. Also, no doubt, we will find ourselves regretfully looking back at many topics that we should have covered more extensively. In any case, however, the most important words are those of Elton, which you will find unedited here.

CHAPTER I: INTRODUCTION

Elton starts his book by defining ecology. Today, Elton's definition of ecology as "scientific natural history" is likely to sound unfamiliar, vague, and even mystifying. Most current texts define ecology as the "science of the inter-relations between living organisms and their environment," a definition that can be traced back Ernst Haeckel (1869). In part the conventional definition may reflect progress on a real intellectual focus about what makes individuals, populations, communities, and ecosystems work. The central ideas of ecology, such as "fitness," "Malthusian parameter," "carrying capacity," "niche," and "ecosystem," all tend to imply a focus on the response of living organisms to their environment, and such concepts often convey the idea that these organisms reciprocally affect their environment. The more modern definition of the field thus seems to focus more on the actual methods and goals of recent ecology.

In contrast, Elton's definition seems less focused. Perhaps this reflected his somewhat iconoclastic hope for the future of ecology in his time. Here and in numerous other places in this book, Elton expresses his conviction that almost any worthwhile observation could potentially provide key insights into solving ecological problems. He even tells us (in Chapter II) that "there is more ecology in the Old Testament or the plays of Shakespeare than in most of the zoological textbooks ever published." Yet he fails to identify what criteria might be applied to consider a natural history observation as being "scientific." By defining ecology so vaguely, Elton may be simply expressing his own enthusiasm, apparent in the rest of this book, for having gained insight from observations of numerous aspects of biology (e.g., diets, behavior, physiology, biogeography, and morphology). The emphasis in this chapter, and in the rest of this book, is on making these observations in the field, or at least in the context of other observations made in the field.

A number of factors have subsequently contributed to the more precise and focused definition we use today. First, consider the improved understanding of evolution. We are told that "nothing in biology makes sense except in the context of evolution" (Dobzhansky 1973), and this is often meant to draw our attention to the importance of adaptive evolution (albeit often under genetic or developmental constraints). In this sense, it is important to realize that natural selection generally describes the "fit between the phenotype of the organism and its environment" and that this relationship is key in protecting the concept of Darwinian evolution from becoming tautological (Brandon 1990). Second, many of the ideas developed by Elton and others at this time (Lotka 1925; Volterra 1928; Gause 1934) ultimately led to the theoretical elucidation of the crucial importance of feedback between organisms and their environment (including other organisms) to our understanding of so many of the basic results of ecology. The conventional definition may thus reflect the perceptual filter that this basic theory puts on natural history. Today, we use these criteria (probably subconsciously) to separate natural history observation into "relevant" and "irrelevant" categories.

An interesting counterpoint is the recent appeal for a "return to natural history" in empirical ecology (Polis 1991; Resetarits and Bernardo 1998). Because recent ecological work has put so much emphasis on experiments, the argument goes, the basic role of "natural history" has too long been neglected. While the appeal of natural history is still very vague, it may reflect the hope that new observations of phenomena, unconditioned by our theoretical expectations, can provide crucial insights into ecology that Elton also seemed to seek. The question might be posed as "is there anything interesting about organisms in the field that can't be evaluated by thinking of mechanisms that describe how organisms interact with their environment?" We won't claim to know the answer to this question, but Elton's oversight of Haeckel's definition may indicate his perspective on this.

Chapter 1 also describes Elton's perspective on how ecology fit into zoology (and biology in general) as a field. He seems particularly defensive about ecology's role relative to the two other major disciplines of biology in his day, taxonomy/systematics and morphology/physiology. Huxley's preface certainly reinforces Elton's goal to establish ecology as a counterpoint, if not an equal, to morphology. While many of these concerns appear dated to us (they certainly don't reflect the importance of genetics that characterizes current biology), they also still resonate with some of the tensions among subdisciplines that characterize ecology today. The tension between "lab-science" and "field-science" is certainly still present and even perhaps exacerbated by two of today's most provocative trends in biology; the advent of molecular (i.e., lab) genetics, and the urgency of environmental (i.e., field) issues related to conservation and ecosystems. Elton's call for closer ties between taxonomists and ecologists is being repeated just as vigorously today (e.g., Wilson 1988).

CHAPTER II:

THE DISTRIBUTION OF ANIMAL COMMUNITIES

This chapter begins with Elton's perspective on defining communities. Unlike the predominant definitions presently being used, Elton's focus was on a definition of communities based on interactions among species, rather than simply delineating communities by area. In doing so, he also hints at the distinction between direct and indirect effects among species, concepts which have received much recent attention (reviewed in Wootton 1994a; Abrams et al. 1996). The discussion of community definitions has waxed and waned throughout the history of ecology; a synthesis of concepts of community and related concepts was recently advanced by Fauth et al. (1996). In contrast to Elton's food web (or "food cycle") emphasis, Fauth et al. base their terminology on a combination of area, taxa, and resource base criteria, with the community being defined exclusively in terms of geographic boundaries. An alternative view more compatible with Elton's interaction-based perspective has been advanced by Cousins (1990), who advocates defining communities by the geographic range of an individual top predator. The contrast between the two approaches represents an ongoing tension between a pragmatic need to delineate the object being studied, which is most easily done using an area-based definition, with a more intellectually satisfying definition based on process. Such process-based definitions may sometimes direct studies in profitable directions; for example, the biological species concept (Mayr 1942) has prompted a focus on reproductive isolating mechanisms. It isn't clear yet, however, whether or not this dialogue over definitions of communities has had similar consequences.

The tension between Elton's perspective and area-based definitions also reflects recent uncertainty about just how integrated a community can be. Such tensions were evident in the debates between Clements and Gleason shortly after Elton wrote his book, and more recently in the debates over the use of null models in community ecology (reviewed by Gotelli and Graves 1996). Defining communities on the basis of interactions automatically presumes that interactions among species are important in shaping community structure, an assumption which is not necessarily true. This question has been inherent in many recent studies of gradients and the concordance of species distributions along them (e.g., Whitaker 1956), a second focus of this chapter. Although there is accruing evidence (especially through experiment) that species interactions are important, recent concerns about the significance of this evidence is focused on the scale at which it acts.

In addition to his concern with the definition of communities, Elton suggests that the study of species distribution along environmental gradients can reveal much about the functional aspects of communities. Elton was clearly impressed by the importance of gradients in many different habitats, gradients which are still the topic of study today, including such factors as depth in oceans and lakes

(Underwood 1978), changing features of flowing waters (e.g., the river continuum hypothesis, Vannote et al. 1980), as well as altitude (Whitaker 1956; Janzen 1967), and latitude (Stevens 1989; Roy et al. 1994). Elton develops this argument along two interesting lines.

First, Elton relates the distribution of species along gradients from the perspective of environmental features. This is the most obvious approach and has since been developed extensively. In the 1950s Bray and Curtis (1957) initiated the first methods for ordination in which the distribution of species among sites could be related to the environmental features of these sites. Since then, a number of elaborations including both direct ordination (with explicit use of environmental data) and indirect ordination (in which distributions themselves are used to infer the relative positions of sites along gradients) have been developed (e.g., Pielou 1984; Jongman et al. 1995).

A more recent trend has been to suggest that the patterns in the positions of species along gradients could reveal the fundamental processes that regulate their populations. Perhaps the best known current example of this approach is that of Tilman (1982, 1988; see also Smith 1983) who suggested that gradients in resource supply could reveal how competition among the consumers of these resources was regulated. Other well developed research programs that use such patterns to infer mechanisms include the effects of disturbance (Ward and Stanford 1983), environmental stress (Bertness and Shumway 1993), productivity (Leibold 1996, 1999; Leibold et al. 1997), and area and isolation (Patterson and Atmar 1986).

Elton emphasizes how continuous environmental gradients can produce sharp discontinuities in the distribution of organisms along these gradients. Interestingly the example he chooses to illustrate this phenomenon involves competition for directionally supplied resources (light versus nutrients in terrestrial plants). His argument essentially points out that plants that are effective at reducing light levels are precisely those whose growth will be most limited by soil nutrients, whereas plants that are good at reducing nutrients will be most limited by light. This occurs because of the architectural consequences of allocation to roots versus shoots as was pointed out by Tilman (1988). However, the consequence of this relationship is that competing pairs of species that differ in allocation will have weaker relative competitive effects on themselves than they do on their interspecific competitors, with the result that alternate stable states are possible while coexistence is not (Reynolds and Pacala 1993). The outcome will depend on which species first gets established, and, as Elton points out, sharp boundaries between species should result.

Elton's second line of argument about inferring mechanisms from the distribution of organisms along gradients seems to take the perspective of the organisms rather than the environment. His arguments here foreshadow recent developments in "macro-ecology" (Brown and Maurer 1989; Brown 1995; Gaston 1999; Gaston and Blackburn 1999), and in understanding the role of organisms' movements across system boundaries (Polis and Hurd 1996). For example, Elton is interested in how generalist and specialist species are distributed among sites, and he points out that locally common organisms tend to have broad ranges elsewhere. While the suggested importance of these patterns is not new, it is only recently that ecologists have started to quantify these patterns and to study them more carefully.

The idea of analyzing species distributions among sites with Elton's joint interest in perspectives taken from the point of species traits (what would be called R-mode analyses in statistical descriptions of these patterns; see Legendre and Legendre 1998) and the point of environmental descriptors (what would be called Q-mode analyses) were perhaps what led Elton to describe the despair of learning about ecology (p. 17). An interesting solution has been recently suggested by Legendre et al. (1997) in which statistical methods are detailed that describe correlations between the traits of species found at a number of sites and the environmental features that characterize those sites.

CHAPTER III: ECOLOGICAL SUCCESSION

Elton explores how animal communities change through time, a natural extension of the intense interest of plant ecologists on ecological succession (especially the work by Tansley). As one of the first organizing principles of community ecology, Elton provides a summary of early ecological succession ideas rather than breaking new conceptual ground. He also follows in the footsteps of Victor Shelford, a student of the plant ecologist and succession pioneer Henry Cowles, by noting that animal succession might passively track plant succession in many cases, but he moves beyond this notion by emphasizing how animals might also actively alter plant succession.

Today, discussion of the process of succession is almost always based on the distinctions that Connell and Slatyer (1977) made between three possible explanations. It is interesting that Elton touches on all three of these ideas with examples in this chapter:

1. Facilitation, or the idea that early colonists modify the environment in ways that eventually give an edge to other species, even possibly leading to exclusion of the early colonists. Elton describes this idea, which had already been well developed by Cowles, Tansley, and Clements (in "stupendous detail" according to Elton), in the introduction of the chapter (p. 20).

2. Tolerance, or the idea that highly dispersing early colonists have little effect on later arriving species that often outcompete or marginalize the early colonists. Elton describes this idea next by discussing how highly dispersing (and rapidly growing) mosses are largely replaced by herbs, shrubs, and trees (p. 21).

3. Inhibition, or the idea that highly dispersing early colonists can resist later arrivals, but that eventually disturbances or grazers differentially disrupt early colonists in ways that allow successional change to occur. In numerous places in this chapter, Elton emphasizes the role that grazers can play in altering succession by both preventing potentially invading species from replacing early species (e.g., vertebrate grazers preventing the replacement of heather moor by pinewoods on pp. 21–22) and by removing plants in ways that allow other species to colonize (e.g., the replacement of heather moor by rushes and docks due to the effects of gulls on pp. 23–24). Interestingly, Elton even emphasizes that it is the *relative* effects of grazers that matter more than their absolute effects (p. 30).

Elton's exposition of succession did not emphasize the distinct predictions and consequences that these three mechanisms yield. Nevertheless, although Connell and Slatyer's ideas have been tremendously important in guiding modern studies of ecological succession, these studies indicate that all three mechanisms are often interwoven in natural systems. For example, Tilman's studies of succession at Cedar Creek indicate that early succession is largely driven by a trade-off between dispersal and competitive ability under repeated disturbances (Tilman 1994, 1997), even though he had earlier invoked a mechanism that was more of a hybrid between facilitation and tolerance (a mechanism that might still explain later stages of successional change).

Similarly, studies on Mt. St. Helens indicate that interactions between early and later colonists can even change from facilitation to inhibition within the course of a succession (Morris and Wood 1989).

Perhaps most striking for its time was the degree to which Elton emphasized that animals might not simply be passive players in ecological succession, but that they can alter successional trajectories through their consumption and dispersal activities. Such ideas were not seriously incorporated into syntheses of succession until Connell and Slatyer's (1977) paper distinguishing facilitation, tolerance, and inhibition provided a clear hypothesis-testing framework which opened the door to experimental tests of succession mechanisms and the role of consumers (e.g., Sousa 1979; Lubchenco 1983; Hils and Vankat 1982). Nevertheless, Elton was not entirely original in appreciating this fact; experimental demonstrations of animals directing plant succession were provided by Tansley and Adamson (1925) and alluded to by Darwin (1859).

A second modern-sounding perspective raised in the book was the idea of succession producing a mosaic of patches at different stages of development across a landscape. Although these ideas are present in some of the early successional works (e.g., Cowles 1899), they were largely ignored by the major treatments of the day (e.g., Clements 1916), and were only emphasized beginning in the 1970s (Dayton 1971; Heinselman 1973; Platt 1975; Denslow 1980; Paine and Levin 1981; Brokaw 1985). Today there is great interest in the role of such patch mosaics in landscape ecology (Pickett and White 1985).

Perhaps most appropriate today is Elton's claim that "succession . . . does not take place with the beautiful simplicity which we could desire, and it is better to realise this fact once and for all rather than try to reduce the whole phenomenon to a set of rules which are always broken in practice!" (p. 27).

Chapter IV: Environmental Factors

Perhaps the most striking discussion in this chapter on environmental factors involves the discussion of the roles of biotic and abiotic factors in regulating the distribution of animals along gradients. To a modern reader, Elton's description of the factors limiting the distribution of *Eurytemora lacinulata* vividly recalls, both in the line of inference from evidence and in their results, the classic work by Connell

(1961) on barnacles. Perhaps most intriguing is the lurking theme of what we would today recognize as the difference between fundamental and realized niches (Hutchinson 1957), although Elton delays in introducing the niche concept until the next chapter. Elton notes that although organisms may tolerate a relatively broad range of physical conditions, their distribution may be more limited than that due to the presence of other competing species within the system. Whether other species always act to limit distributions is debatable, since the presence of resource species may provide opportunities for focal species to persist in physically stressful environments. In some of Elton's examples, the limitation appears to arise through the direct consequences of another species on the biotic or abiotic environment. However, in other cases, the limitation appears to arise from adaptive behavior, foreshadowing more recent work distinguishing proximate and ultimate factors (Orians 1962), and developing habitat selection theory (Fretwell 1972; Rosenzweig 1981).

On a more general basis, this chapter (along with the next chapter) is focused on "what animals do," a phrase that Elton repeatedly emphasizes but never fully explains. In this chapter the focus is on environmental factors and especially on identifying which of these factors is important, especially in the form of "limiting factors." Even though Elton doesn't cite von Liebig's "law of the minimum" (1855), his discussion is strongly favored by the idea that despite the huge number of potentially important factors that potentially influence a population in a given environment, usually "just one or two factors" are limiting in that they determine whether the species can be present or not at that site. It is also unlikely that Elton wasn't aware of von Liebig's work, especially since he was probably familiar with Blackman who was an early advocate of von Liebig's law. Interesting is the fact that Elton was a bit less dogmatic than von Liebig (see American Association for the Advancement of Science 1942 for an early history of von Liebig's law) by allowing for more than one limiting factor, but still emphasizing that the number of factors should be very small. He also emphasized how this focus on a small number of limiting factors could greatly simplify the analysis of species distributions by revealing what prevents species from "doing the things that they do."

Von Liebig's "law of the minimum" has long been important in ecology and is still an important feature (or outcome) of many population models. For example, the recent models of resource competition developed by Tilman (1982) specifically model the growth of resource-limited consumers as being governed by von Liebig's law; if resource ratios differ from some threshold value, growth is limited with a Michaelis-Menten dynamic based only on the more limiting resource. In contrast, animals are typically limited by multiple resources which are substitutible or complementary (*sensu* Tilman 1982), hence many animal ecologists are much less influenced by von Liebig (recent interests in "stoichiometric" theories of animal ecology may represent an interesting exception; see Elser et al. 1988; Sterner 1994).

Nevertheless, thinking about the relevance of von Liebig's law continues to develop. First, it is important to realize that the concept of "limitation" under this view is relevant to the short-term time scale (see Osenberg and Mittelbach 1996 for a discussion of limitation and regulation at different time scales). For example, it is commonly the case that joint additions of nutrients increase plant densities much more than the addition of any single nutrient, even though there is usually one nutrient that is much more effective than the others when added individually. One explanation is that as soon as one limiting factor is removed, another one rapidly takes its place (Elser et al. 1988), a view von Liebig espoused (American Association for the Advancement of Science 1942).

Another more complex and sophisticated argument which can explain this observation is based on the idea that organisms should adjust their allocation strategies so as to be jointly limited by different resources (Gleeson and Tilman 1992). This latter argument about the behavior of systems under joint evolutionary (or at least adaptive) and population dynamic constraints can lead to greatly modified interpretations of limiting factors (Chapin et al. 1987; Gleeson and Tilman 1992). Finally, when limiting factors are viewed as stressors on organisms, rather than as resources, it is easy to envision that factors interact to limit populations (such as joint impacts of osmotic stress and competition with other species in Elton's *Eurytemora* example).

The relevance of adaptation (in its broadest sense to include adaptive phenotypic plasticity as well as adaptive evolutionary change) as an important assumption in interpreting patterns of animal distribution also seems to have been appreciated by Elton in what again appears to be a remarkably modern (though perhaps still very controversial) way. Especially intriguing is Elton's suggestion that by assuming optimized behavior (especially habitat selection), one can learn much about how animals might be limited by different environmental factors, even when this limitation is not revealed by variation in demography. This seems to imply some sense that animals might make these decisions based on marginal value comparisons among multiple choices in habitat selection. Current theory in optimal foraging behavior theory is greatly driven by the analysis of such decisions (see Stephens and Krebs 1986) and especially in the complex problems associated with decisions based on multiple currencies, such as finding compromises between risk and energy (see Caraco 1980; Real 1980). The most extreme development of a research program along these lines today is perhaps exemplified by the theoretical work of Cohen et al. (1991). Elton would perhaps be surprised by the level of controversy of this approach (due to its emphasis on optimization and ESS approaches at the perceived expense of realism in the constraints that might prevent such optimization), but would probably also be surprised by the range of issues that can be addressed with such an approach. Others who have contrasted purely population dynamic models with those that include the optimization of behavior include Abrams (1992), Sih (1992), and Holt (1996), and some model systems for these effects include fishes (Werner and Gilliam 1984; Gilliam and Fraser 1987), amphibians (Sih et al. 1988; Werner and Anholt 1996), zooplankton (Leibold and Tessier 1991), rodents (Brown et al. 1994), and insects (McPeek, 1998).

Chapter V: The Animal Community

This chapter contains the core ideas of Elton's perspective on ecology. Upon initially reading this chapter, one is amazed both by the degree to which modern topics are identified and the density of conceptual ideas. It is only upon rereading the chapter that one begins to appreciate the intimate interconnections Elton sees among the different topics.

The chapter begins with a discussion of indirect effects, which Elton first mentions in Chapter II (p. 12). His discussion highlights the unexpected consequences possible when indirect effects are present, mirroring the renewed attention ecologists have given indirect effects in the past several decades as the recent wave of experimental manipulations have routinely produced surprising results that are only explicable in this context (e.g., Martin et al. 1992; reviewed in Wootton 1994a). In this context, Elton also reminds us that humans are intimately connected to ecological communities, identifying the important point, largely forgotten by modern biomedical science, that disease dynamics depend on their ecological context (e.g., Jones et al. 1998). The problem of indirect effects is a very important one for community ecology since it can be shown that the effect of any species on any other species can depend critically on the exact composition of the rest of the community, a problem that Schaffer (1981) has called the problem of "ecological abstraction." This means that net effects (those mediated through the cumulative effects of all direct and indirect pathways) of species on each other (and on other features of the ecosystem) can be complex functions of their direct effects and of those of the other species with which they coexist. Methods for evaluating such net effects include loop analysis (Levins 1975), the use of the "inverted community matrix" (Levine 1976), and the closely related derivation of the "gamma matrix" (Lawlor 1979). Empirically, the study of indirect effects in dynamic systems is still a problematic issue; some have suggested that path analysis can be helpful (Wootton 1994b), but this is still controversial (Petraitis et al. 1996; Smith et al. 1997; but see Grace and Pugesek 1998; Pugesek and Grace 1998). Others have tried to evaluate loop analysis (Lane and Levins 1977; McCauley and Briand 1979), and the inverse matrix approach (Levitan 1987; Schmitz 1997) with varying degrees of success. Finally, the presence of indirect effects has also often been approached using methods related to ANOVA and other statistics (Wilbur 1987; Wilbur and Fauth 1990; Miller 1994; Menge 1995; Morin 1995).

As an obvious attempt to simplify the possible range of species interactions that might be considered, Elton suggests that many of the interactions leading to indirect effects revolve around consumer-resource interactions, and that such interactions are fundamental to all ecosystems, a point which seemed inexplicably forgotten in the community ecology of the 1960s and 1970s when the focus was on competition coefficients that didn't explicitly consider resources. This focus on trophic interactions allowed Elton to better develop the four central concepts of this chapter. Some recent attempts at reintroducing nontrophic interactions into thinking about communities include the concept of the "interaction matrix" (Menge and Farrell 1989) and the concept of "ecosystem engineers" (Jones et al. 1994).

The first of Elton's central concepts is that of food chains and food cycles (which we today call food webs). Today we understand that the dynamics of food chains and food webs are quite distinct (see Leibold 1989, 1996; Hunter and Price 1992; Power 1992). Elton argues that food webs provide a concepual framework to understand species interactions in the context of complex ecosystems. Although earlier workers had generated food web diagrams as an aid to understanding ecological systems (e.g., Camerano 1880; Shelford 1913), Elton substantially advanced their use by proposing that they play a central organizing role in ecology. This position subsequently has inspired a huge amount of work on food webs along several lines (unfortunately not all of which seem to be well integrated). Some have developed sophisticated metrics to describe the features of food web diagrams (such as Elton's Figures 3 and 4). While not all agree that these metrics are meaningful (Paine 1988; Polis 1991), this work has served to identify possible variation and regularities in food web architecture (see Cohen and Briand 1984; Martinez 1994), which may have important relations to system stability and other attributes (e.g., Pimm 1979, 1982). Others have investigated species interactions in food webs (e.g., Paine 1966, 1980; Neill 1974, Spiller and Schoener 1990; Wootton 1992, 1994b; Leibold and Tessier 1998; see also papers in Polis and Winemiller 1996) in an attempt to better quantify and illustrate how interactions have ramifying effects in food webs.

Elton then turns his attention to the role of size in ecological interactions arguing for its central role in organizing the structure of food webs. In doing so, he again touches on several modern themes, including the allometry of life history traits (e.g., Peters 1983; Harvey and Zammuto 1985; Wootton 1987), and ecological dynamics (Damuth 1981; Gaston and Lawton 1988; Brown 1995). Size relationships also form the basis for some modern theories of food web architecture (Cohen 1989), of ecosystem dynamics (Pahl-Wostl 1995), and of social behavior's role in overcoming size constraints in consumer resource interactions (Packer et al. 1990). Nevertheless, the promise of using size as an index of ecological function, as an efficient way to parameterize food web models for example, has not yet reached fruition.

This chapter also holds a prominent place in the discipline as one of two sources that introduced the niche concept into ecology. Though it isn't exactly clear just what Elton intended by his "role of the species" definition (see discussions in Schoener 1989; Leibold 1995), his use of the term in this book clearly led to its greater use (e.g., Allee et al. 1949). His use of the term seems to emphasize a "functional" view based on the position of species in a food web whereas Grinnell (1917) seems to emphasize the requirements of species (an approach that led to Hutchinson's [1959] formal definition). Ironically, MacArthur and Levins (1967) then developed "niche theory" with an approach much more closely related to Elton's roles than to Hutchinson's definition, even though Hutchinson was MacArthur's advisor. Since then, "niche theory" has been criticized on both empirical and conceptual grounds (e.g., papers in Strong et al. 1984), and the use of the word "niche" has greatly declined in recent years. Nevertheless, we suspect that the concept is still critical in the informal reasoning of ecologists, and it might be a mistake to abandon it. Given the increased "mechanism" that characterizes the study of species interactions today, it is clear, however, that the niche concept as originally conceived needs to be substantially expanded, perhaps through a synthesis of the functional and requirement-based perspectives (Leibold 1995).

The final concept Elton introduces is the "pyramid of numbers." One of the explanations for this pyramid of numbers involved the idea that organisms are smaller at lower trophic levels (e.g., Elton's ideas about size relations mentioned above) but there is also often a trophic biomass pyramid (Odum 1953), and this had led to a more general discussion of "trophic pyramids" in ecosystems. In his work, Lindeman (1942) synthesized Elton's concept with Lotka's (1925) energetic perspective, which launched the field of ecosystem ecology, and shifted the focus to patterns of biomass rather than numbers. Elton's ideas are strongly based on size considerations, such as lower per-individual requirements and higher reproductive rates of small individuals, but miss an important component of more modern thinking: the inefficiency of energy transfer. Recent discussions have also noted that pyramids of biomass, if not numbers, can be inverted in some circumstances, particularly in aquatic systems (e.g., Piontkovski et al. 1995; Buck et al. 1996). Ironically, one of the mechanisms invoked to explain inverted pyramids, rapid turnover of basal species in food webs, requires the high reproductive rates Elton identifies as causes of pyramids. The regulation of trophic structure (relative bio-mass patterns among organisms in different trophic levels or functional groups) is one of the most active areas of research at the interface of community and ecosystems ecology because it is unclear whether either food chain models of trophic interactions (Power 1992) or more complex food web interactions (Phillips 1974; Abrams 1993; Leibold 1996) explain natural patterns of trophic structure very well. One possibility is that shifting composition of food webs either through variation in the number of trophic levels (Oksanen et al. 1981), via shifting importance of omnivory (Polis and Holt 1992) or through shifting composition of species within trophic level (Leibold 1996) is important in the regulation of trophic structure. As suggested by Elton, variation in trophic structure and in the shape of the trophic pyramid also implies variation in the rates of energy and material fluxes in food webs (Odum 1953; Baird and Ulanowicz 1989; Higashi et al. 1993; Bondavalli and Ulanowicz 1999). It is clear that a major pattern contrasts typical terrestrial ecosystems with aquatic ones (Hairston and Hairston 1997) but other patterns may also be present.

It is nevertheless the case that, despite 80 years of work on issues related to these four central themes, the synthesis envisioned by Elton has yet to be attained.

Chapter VI: Parasites

In this chapter, Elton extends the themes of Chapter V to encom-pass parasites and pathogens. Then, as is perhaps still true, diseases and parasites received much less attention in conventional ecological thinking as did predators. Nevertheless, Elton strives to incorporate parasites into the same fundamental themes of Chapter V, including size relationships among consumers and their resources (inverted in this case), the pyramid of numbers (again inverted), niche relations, and food web structure. In contrast with discussions of predators, however, his comments give the impression that the effects of para-sites on their hosts is much less than the effects of predators on their prey. He also mentions inefficiency in trophic transfers as an impor-tant factor limiting the length of parasite food chains, even though this concept is conspicuously absent in his treatment of predators.

In part Elton's view of parasites as being under "donor control" seems related to his view of parasites as being involved in an "elaborate compromise between extracting sufficient nourishment . . . and not impairing . . . its host, which is providing it with . . . a free ride" (p. 72). This statement has very modern associations with many of the current epidemiological models of disease ecology (Anderson and May 1985) which harken back to the Kermack and McKendrick models published in 1927 (interestingly Elton doesn't cite this paper either here or in later work), which were originally motivated by situations in which pathogens induce immunity in their hosts, rather than causing death or appreciably reducing fecundity.

The extent to which Elton discounted the effects of parasites on food webs is interesting, despite his call to treat parasites "essentially the same as carnivores," and the fact that there were then already some well known successes in the use of parasites and other enemies in biocontrol. For example, the control of the citrus cottony cushion scale (*Icera purchasi*) was achieved in 1889 by A. Koebele (the "father" of biocontrol [Price 1975]). Although the successful agent in that case was a predator (the coccinelid *Rodolia*), early efforts were more focused on a parasitic fly (*Cryptochaetum*) apparently under the general impression that parasitoids were better biocontrol agents than predators (Price 1975). The study and use of parasitoids in pest control, including work by Nicholson and Bailey (1935), Varley (1947), and others, increased dramatically shortly after Elton's book was published.

Today the importance of diseases and parasites is much better understood and even appreciated by ecologists even though most of the attention by field ecologists remains focused on predators. Nevertheless, the role of diseases is increasingly recognized even in habitats where they had once been essentially ignored. A striking recent example is the discovery of extremely high densities of viruses in oceanic ecosystems and the recognition that they might play crucial roles in biogeochemistry, ecosystem processes and community organization (Fuhrman 1999).

Finally, Elton draws a distinction between parasite hosts and carnivore prey in that the former may serve a dual role as both food and agents of dispersal. Such commensalistic (and related mutualistic) interactions are another group of interactions which have historically received less attention by ecologists, and they are barely discussed in this book, despite their obvious importance in many instances (e.g., plant-pollinator interactions).

CHAPTER VII: TIME AND ANIMAL COMMUNITIES

Temporal fluctuations and temporal cycles are factors that have always made the life of ecologists more difficult. Here Elton focuses on environmentally-driven fluctuations such as diel, tidal, weather, and seasonal cycles (leaving other, less predictable fluctuations for chapter IX). As any good naturalist will attest, these and other cycles play obvious roles in the ecology of organisms. Nevertheless, the existence and consequences of more subtle cyclical fluctuations are still being uncovered, such as the recent interest in El Niño events (Glynn 1988), "interdecadal" oceanic fluctuations (Mantua et al. 1997; Downton and Miller 1998; McGowan et al. 1998), and the legacy of glacial and interglacial periods (Davis 1986). In addition to describing the effects of these cycles on the physiology and behavior of organisms, Elton pushes the idea that these cycles provide opportunities for niche differentiation and even for the existence of "convergent" communities during different periods of these cyclical fluctuations.

The roles of such cycles on the physiology and behavior of organisms has since received tremendous attention. For example, the factors that regulate seasonal bird migration which were poorly known then are now understood to be a complex array of hormonal, physiological, sensory, and social networks of interacting factors. Rather than a simple "bottom-up" perspective in which resources can be partitioned by these temporal cycles into day/night or winter/summer dichotomies, more recent work has emphasized both the roles of multiple factors and the flexibility with which organisms can change their activity levels in response to variation in such factors.

In addition to resources, activity cycles of organisms are now understood to respond to factors including predation risk, and could also easily involve other factors. Even in cases where physiological considerations such as those developed by Elton are operating, the more ultimate consequences of these physiological effects are often on other factors such as avoiding predation (e.g., lizards and migrating birds). The regulation of diel cycles in activity by predation is obvious even in organisms with relatively simple behavioral repertoires such as zooplankton (Zaret 1980; Leibold and Tessier 1998) in addition to organisms with greater behavioral flexibility such as desert rodents (Kotler et al. 1993; Brown et al. 1994).

Elton focuses on cases where parallel trophic links seem to have evolved, such as those involving bats and moths versus swallows and butterflies, to suggest that these result from partitioning resources in time due to contrasting temperature and light (as well as other factors). Whether this represents an important mechanism for niche partitioning remains unclear; for example, if diurnal birds and nocturnal rodents both consume seeds, such temporal partitioning of activity will have minimal effects unless they are feeding on very distinct seeds as well. Elton's notion of temporal partitioning in the importance of food chains seems reasonable, but it implicitly assumes that consumption is predicated on encounters among active organisms. If consumers are capable of finding quiescent organisms (e.g., bears consuming beehives at night or digging up hibernating ground squirrels), however, such a distinction would again break down.

Taking the idea of niche partitioning via these periodic fluctuations even further, Elton also suggests that parallel food chain and food web architectures may often evolve. He does, however, recognize that convergence of food web structure during different periods of cyclical change will not always be exact or even occur at all. However, the degree to which his hypothesis is true or generalizable is probably impossible to evaluate. Even cases where convergence of communities (not involving such purely temporal bases) have been most apparent (e.g., comparisons of Sonoran and Negevian rodent assemblages), the evidence that identical patterns of species interactions, such as those hypothesized by Elton, are at play is certainly not convincing (Brown et al. 1994). We are not aware of any studies that have followed up Elton's hypothesis by comparing food web architectures during different diel or seasonal periods, even though accounting for such variation is one of the criticisms leveled at many current food web studies (Paine 1988).

CHAPTER VIII: THE NUMBERS OF ANIMALS

In this and the following two chapters, Elton explores central issues in what we now characterize as population ecology. Although the conceptual impact of *Animal Ecology* was most strongly felt at community and ecosystem levels of organization, the subsequent career of Elton and his students focused more on the themes raised in these chapters, which he characterized as "in an extremely early stage . . . but . . . of profound importance" (p. 102). The chapter also raises two very modern applied themes: the roles of humans in causing extinctions and in facilitating exotic species invasions.

Initially, the chapter focuses on the actual and potential abundances of animals in nature. While Elton mostly painted a picture of overall population size and its potential to increase, subsequent studies have been inspired to explore more systematic relationships with population size, including Preston's (1962) characterization of abundance distributions, and in the relationship between abundance and body size (e.g., Van Valen 1973; Damuth 1981; Brown 1995), a nascent theme of Elton's. An interesting aspect of Elton's discussion is his treatment of extinct and endangered species here. Although he attributes some extinction "to the greed of individual pirates," his view expresses a rather cavalier attitude by the standards of today's ecologists about the prospect of species extinction: "It is not much use mourning the loss of these animals, since it was inevitable that many of them would not survive the close settlement of their countries" (p. 106). In retrospect he may have been too pessimistic, since he saw bison and whales as doomed to extinction; subsequent protection efforts have led to substantial recovery, and an entire field of conservation biology is predicated on the possibility that extinction from human activity is not necessarily inevitable.

Elton also uses the dynamics of exotic introductions to illustrate his view that tremendous potential population growth rates are ultimately brought under control by food web interactions. Here the discussion of this very modern topic is limited to illustrating the tremendous population growth rates of species and to describing a generalized scenario for these introductions involving fast early growth and subsequent control by enemies involving adjustments by their predators. Elton followed up on these general themes in his important 1958 book on invasions.

Elton then turns to the issue of how populations are regulated, a central preoccupation of ecology through most of its subsequent history. In some ways, Elton presents a fairly modern tone, taking a community-level perspective featuring the interplay between competition for resources and control by consumers. Elton argues that the counterbalancing effects of food and predators are important in regulating abundances within bounds, preventing species from becoming so rare that they would go extinct via Allee effects (not described as such here [see Allee 1931]) or from catastrophic overexploitation of their food. This "balancing" perspective differs from the influencial one developed by Hairston et al. (1960) which was premised on the action of one or the other (depending on trophic level) but not both and is more closely related to the more modern perspectives (Oksanen et al. 1981; Carpenter et al. 1985; Leibold 1989, 1996; Hunter and Price 1992; Wootton and Power 1993).

Elton struggles a bit while discussing the regulation of top carnivores. Although some mechanisms, such as interference, are now considered generally important, his discussions of species maintaining "optimum" population sizes are similar to more recent ideas of "prudent predators" (Hart et al. 1991). Such mechanisms often require group selection, which is currently not thought to play a generally important role in evolution.

Embedded in Elton's discussion of population regulation are several other themes of modern interest. For example, he illustrates the difficulty of determining the mechanisms and directional effects of species in controlling other populations because of indirect effects, a major motivation for recent studies attempting to evaluate the strengths of interactions (e.g., Paine 1992; Laska and Wootton 1998; Berlow et al. 1999). He also considers the effects of omnivory, using examples of what have garnered recent interest under the name of "intraguild predation" (Polis and Holt 1992). He also proposes that omnivory, by reducing the risk that a consumer will become extinct via starvation, acts to stabilize food webs, a proposition reiterated by MacArthur (1955). This hypothesis does not appear to hold in models of food webs with linear per-capita interactions (Pimm and Lawton 1978), although recent studies of nonlinear systems (McCann et al. 1998) suggest that it may indeed be correct under some realistic circumstances. A yet unexplored hypothesis which may merit study is the related proposition that omnivory increases with higher trophic position in the food web.

CHAPTER IX:

VARIATIONS IN THE NUMBERS OF ANIMALS

This chapter gives a counterpoint to the previous chapter's focus on tight regulation of animal numbers. From Elton's naturalist perspective the fact that many populations fluctuate greatly cannot be ignored despite his conviction of tight regulatory processes in nature, and his research group spent considerable energy on the problem in subsequent years. Still, the contrast in perspective between the two chapters seems to foreshadow the intense debates over the role of density-dependent regulation of populations (Nicholson and Bailey 1935; Andrewartha and Birch 1954; Lack 1954; Strong 1986).

After presenting a number of examples of fluctuating populations of animals, Elton turns his attention to the possible causes of these fluctuations. His emphasis is on external perturbations of the physical environment leading to loss of regulatory control by predators, with disease or parasites eventually bringing a halt to the outbreak. Such a pattern indeed seems to be correct in some recent studies of fluctuating populations (e.g., the gypsy moth in eastern North America; Elkinton et al. 1996; Jones et al. 1998). One aspect of his argument, which is not well developed, is why many of the examples presented involve regular fluctuations. Beyond sunspots, compelling cyclical physical mechanisms leading to regular fluctuations are lacking. This circumstance raises the curious issue of why, despite the emphasis on trophic interactions as an organizing theme, Elton did not invoke consumer-resource interactions in driving cycles, given the prior publication of Volterra's classic paper addressing precisely this issue. Perhaps this omission can be traced to Elton's distrust of ecological theory. Elton read Lotka's work and had this to say: "Like most mathematicians he takes the hopeful biologist to the edge of a pond, points out that a good swim will help his work, and then pushes him in and leaves him to drown" (Elton 1935). Elton does touch on some themes that have subsequently been important in predisposing consumer-resource interactions to cycle, such as time delays in consumer populations and consumer satiation, but he overlooks the possibility that consumers can eventually catch up and overexploit their resources, an important component to driving consumer-resource cycles. Perhaps he was too wed to the idea of prudent predators espoused in the previous chapter.

Elton ends the chapter by considering some implications of population fluctuations, admitting that he only scratches the surface. He chooses to focus on the implications of diet patterns in organisms—emphasizing the effects of both quantity and quality on diet breadth—and ultimately on habitat choice. In many ways his treatment anticipates the early analyses of optimal foraging theory (Emlen 1966;

MacArthur and Pianka 1966) on diet composition, and he sends a clear warning about the consequences to food web diagrams of such diet shifts, a warning which is quite germane to the studies of static food web architecture prevalent today (Cohen and Briand 1984; Martinez 1994).

In the years since Elton's book the implications of fluctuating populations have been developed in several important directions, and continue to be an exciting area of investigation today. The revelation that chaotic dynamics could be produced from simple ecological models with strong density-dependence and time lags (May and Oster 1976) has launched intensive efforts to identify whether such population fluctuations arise from chaos rather than external variability in the physical environment (e.g., Hastings et al. 1993; Ellner and Turchin 1995). Population fluctuations have also been suggested as an important mechanism that might promote coexistence among species, starting with Hutchinson's "paradox of the plankton," more recent investigations of the storage effect (Chesson and Huntley 1997), and nonequilibrium dynamics arising from nonlinearity (Armstrong and McGehee 1976; Levins 1979). Another wrinkle on this theme has been the investigation of variability's effects within stage-structured populations on subsequent population dynamics and coexistence (Roughgarden et al. 1988; Nisbet et al. 1989; Pfister 1996, 1998). These and other issues concerning population fluctuations are far from settled, and we can expect further advances in the future.

CHAPTER X: DISPERSAL

Elton recognized the importance of dispersal to ecology in this chapter and noted the difficulty of characterizing it. Dispersal remains a critical issue in ecology today, and, despite technological innovations such as elemental fingerprinting (Swearer et al. 1999; DiBacco and Levin 2000), the study of dispersal remains daunting.

Elton's interest focused on species colonization of new areas, a theme elaborated in more detail in his later book on invasions. His separation of the colonization process into dispersal, establishment of the individual, and establishment of the species provides a useful framework for organizing ideas and executing a research strategy on the colonization process (although the addition of another category, focused on the dynamics of spread, would usefully describe many recent studies). Elton's approach is fairly anecdotal in this chapter. More recent studies have emphasized quantitative approaches, facilitated by the development of mathematical theory, such as diffusion and telegraph equations (Skellam 1951; Okubo 1980; Holmes 1993). Such theory has served to sharpen the focus of dispersal studies on critical parameters to measure, and has led to more sophisticated analyses of the dynamics of colonization in several systems (sea otters in Lubina and Levin 1988; marine invertebrates in Grosholz 1996; house finches in Veit and Lewis 1996; pest insects and their pathogens in Dwyer and Elkinton 1995). Still, predictions of species spread as a function of dispersal turn out to be highly sensitive to the shape of dispersal functions, particularly at the extreme tails (Lewis 1997), and these shapes are difficult to distinguish empirically. Elton's call for observations of long-distance dispersal events seems especially important given today's modeling results.

More recently, documenting dispersal rates and understanding their implications have become the focus of intensified investigation with an appreciation of spatial subdivision and fragmentation of populations and communities—a feature that Elton barely alludes to in his discussion of oxbow lakes. Beginning with Andrewartha and Birch's (1954) views on population dynamics and Levins's (1969) conception of metapopulations, there has been renewed interest in the role dispersal plays in maintaining subdivided populations (Ehrlich et al. 1975; reviews in Gotelli 1991; Hanski and Gilpin 1997), and this theory has become central to the field of conservation biology in light of habitat fragmentation. Quantifying dispersal patterns and their relationship to extinction is a central requirement of successfully implementing metapopulation perspectives into empirical settings.

Beginning with Huffaker's (1958) laboratory experiments with mites, the implications for dispersal in subdivided multispecies systems has also been developed, and remains an active area of investigation. MacArthur and Wilson's (1969) theory of island biogeography considered the role of dispersal in establishing patterns of species diversity in subdivided habitat. Building on Huffaker's results, recent studies have established theoretically (e.g., Tilman 1994; Loreau and Mouquet 1999) and empirically (Root 1973; Kareiva 1987) that dispersal limitation can be critical in promoting species coexistence by effectively weakening strong interactions among species. The implications of dispersal in conjunction with spatial subdivision, however, still need to be worked out more extensively in multispecies systems, and this area promises to be another area of active investigation in the future.

CHAPTER XI: ECOLOGICAL METHODS

Most of Elton's book is conceptually oriented and is directed at the development of ideas. However, as Elton points out in this chapter, the theories and concepts of a field play a very large role in determining the methods that go about evaluating these theories. This chapter reads a bit like a proscription to students about how to do field work directed at Elton's ideas about the central role of food webs. Much of this advice is still valuable today including the need for precision (especially in the identification of species), and the need for care to avoiding the selective emphasis only on facts that are consistent with preconceived theories.

However, as the field of ecology has progressed, methodological concerns have evolved tremendously to include the technical aspects of measuring the distribution of organisms, as well as measuring the processes that affect their populations. In addition, there has been a tremendous literature on statistical issues that would perhaps have surprised Elton at the time he wrote this book.

Elton viewed the use of surveys and observational methods as essential to the documentation and interpretation of species interactions. His cautions sound remarkably modern in emphasizing the variation that can occur in species' feeding habits, as well as the need to identify feeding interactions at fine scales of taxonomic resolution. He also mentions the importance of quantitative data in supplementing the qualitative information.

One of the most striking recent methodological issues in community ecology is related to the use of experiments. Although Elton mentions experimental methods in this chapter, they certainly don't hold the central position that experiments play in modern community ecology. Remarkably however, modern community ecologists are realizing that experiments aren't everything and are increasingly promoting a more synthetic approach that compares experimental results to correlational studies such as surveys (see papers in Resetarits and Bernardo 1998, for example).

In the future, ecologists may want to consider how they, like Elton, can profitably synthesize natural history with conceptually general approaches in ecology. Recent attempts to more effectively grapple with indirect effects have established ways to insert natural history information into statistical or theoretical analyses. For example, using structural equation modeling to uncover plausible pathways of indirect effects requires *a priori* hypotheses of how species interact with each other, derived from natural history observations (Wootton 1994b; Grace and Pugesek 1997). In addition, the recent development of mechanistic, rather than mathematical, criteria to investigate whether the species composition of an ecological community modifies the interaction between individuals of different species allows a substantial role for natural history observation (Abrams 1983; Wootton 1993). Finally, because we believe that natural history intuition is often a very effective guide to uncovering how different ecosystems function, an intriguing line of inquiry that ecologists may consider in developing new methodologies is to ask the question, "What general features underlie natural history intuition and how might these be quantified and exploited?"

While Elton was a great ecologist, most would probably not list him among the best evolutionary biologists. This chapter reveals some interesting perspectives of his view of evolution, but to our knowledge, Elton's ideas played little role in the development of modern evolutionary theory. Clearly, Elton's objections to Darwin's (1859) theory of evolution by natural selection seem naïve today. It is worth remembering, however, that the modern synthesis was achieved after this book was published (Fisher 1930; Dobzhanski 1937; Mayr 1942; Simpson 1944). What is more striking is the strong prose that Elton used to make his arguments.

Elton focused on what he called the "species problem," which he described as the fact that sibling species often show differences that are not apparently adaptive, whereas more distantly related taxa show differences in traits that seem to have more apparently evolved in response to selection. The evidence for his arguments is today better understood. We know that color polymorphisms such as those Elton described for the fox and the cob have evolved under more complex evolutionary scenarios than he envisaged, including, for example, sexual selection. An important resolution of Elton's "species problem" may come from a better understanding of the complex interplay between sexual and context-dependent patterns of natural selection. Since Elton apparently had mixed feelings about his early relations with Julian Huxley, one might conjecture about the reasons he might have had for making such strong critiques of the Darwinian paradigm.

Attitudes about the relevance of evolution to ecological studies has varied tremendously in ecology. There is a strong sense among some scientists that the two are so closely interwoven that neither can be studied in isolation. Alternatively, one is surprised at how often basic ideas about species interactions are developed in the absence of evolutionary considerations. In the fields of population ecology and ecosystems ecology, as well as in community ecology, many principal tenets have little to do with evolutionary issues. This is apparent in models that are only concerned with the equilibrium behavior and stability of dynamic equations, but that ignore questions about invasibility or evolution on these results.

The difference among the results of theory focused only on dynamic equations but lacking in evolutionary concerns is most strongly developed in theories of "adaptive dynamics" (e.g., Abrams 1992; Cohen et al. 1999). This body of work illustrates how the outcome of species interactions in the presence of adaptation (and especially co-adaptation) is substantially modified from outcomes expected in the absence of adaptations (Abrams 1992).

Of course, there is some debate about how much local adaptation there is in nature. If local populations do not respond adaptively to local environmental conditions, the purely ecological models are probably a good start. If local populations respond optimally to local environmental conditions, the newer "adaptive dynamics" models are probably better. But what happens when there is some intermediate level of local adaptation? Most of the work done on the evolutionary dynamics of ecological traits in natural populations indicates that populations can and do respond to selection, but that this response is often constrained by a variety of genetic, environmental, and demographic factors. Though we now know much, much more than we did in Elton's time, we still don't have much of a sense of what happens when there is adaptive but not optimal evolutionary change in local populations.

Conclusion

Elton's final comments are directed at encouraging fieldwork, emphasizing how an ecologist is often balancing the search for simplifying theories with the recognition of complexity in nature. It is worth keeping in mind that this balancing act will often lead to rhetoric and disagreement among ecologists.

References

Abrams, P. A. 1983. Arguments in favor of higher order interactions. *American Naturalist* 121:887–91.

———. 1984. Foraging time optimization and interactions in food webs. *Am. Nat.* 124:80–96.

———. 1992. Predators that benefit prey and prey that harm predators: Unusual effects of interacting foraging adaptations. *Am. Nat.* 140:573–600.

———. 1993. Effect of increased productivity on the abundances of trophic levels. *Am. Nat.* 141:351–71.

Abrams, P., B. A. Menge, G. G. Mittelbach, D. Spiller, and P. Yodzis. 1996. The role of indirect effects in food webs. In *Food webs: Integration of patterns and dynamics*, ed. G. A. Polis and K. O. Winemiller. 371–95. New York: Chapman & Hall.

Allee, W. C. 1931. *Animal aggregations: A study in general sociology*. Chicago: Univ. of Chicago Press.

Allee, W. C., A. E. Emerson, O. Park, T. Park, and K. P. Schmidt. 1949. *Principles of animal ecology*. Chicago: Univ. of Chicago Press.

American Association for the Advancement of Science. 1942. *Liebig and after Liebig; a century of progress in agricultural chemistry*. Washington D.C.: American Association for the Advancement of Science.

Anderson, R. M., and R. M. May. 1985. Vaccination and herd immunity to infectious disease. *Nature* 315:323–29.

Andrewartha, H. G., and L. C. Birch. 1954. *The distribution and abundance of animals*. Chicago: Univ. of Chicago Press.

Armstrong, R. A., and R. McGehee. 1976. Coexistence of species competing for shared resources. *Theoretical Population Biology* 9:317–28.

Baird, D., and R. E. Ulanowicz. 1989. The seasonal dynamics of the Chesapeake Bay USA ecosystem. *Ecological Monographs* 59:329–64.

Berlow, E. L., S. A. Navarrete, C. J. Briggs, M. E. Power, and B. A. Menge. 1999. Quantifying variation in the strengths of species interactions. *Ecology*. 80: 2206–24.

Bertness, M. D., and S. W. Shumway. 1993. Competition and facilitation in marsh plants. *Am. Nat.* 142:718–24.

BONDAVALLI, C., AND R. E. ULANOWICZ. 1999. Unexpected effects of predators upon their prey: The case of the American alligator. *Ecosystems* 2:49–63.

BRANDON, R. N. 1990. *Adaptation and environment.* Princeton: Princeton Univ. Press.

BRAY, J. R., AND J. T. CURTIS. 1957. An ordination of the upland forest communities of southern Wisconsin. *Ecol. Monog.* 27:325–49.

BROKAW, N. V. L. 1985. Gap-phase regeneration in a tropical forest. *Ecology* 66:682–87.

BROWN, J. H. 1995. *Macroecology.* Chicago: Univ. of Chicago Press.

BROWN, J. H., AND B. A. MAURER. 1989. Macroecology: The division of food and space among species on continents. *Science* 243:1145–50.

BROWN, J. S., B. P. KOTLER, AND W. A. MITCHELL. 1994. Foraging theory, patch use, and the structure of a Negev Desert granivore community. *Ecol.* 75: 2286–300.

BUCK, K. R., F. P. CHAVEZ, AND L. CAMPBELL. 1996. Basin-wide distributions of living carbon components and the inverted trophic pyramid of the central gyre of the North Atlantic Ocean, summer 1993. *Aquatic Microbial Ecology.* 10:283–98.

CAMERANO, L. 1880. On the equilibrium of living beings by means of reciprocal destruction. Translated by J. E. Cohen in *Frontiers in mathematial biology*, ed. S. A. Levin. New York: Springer-Verlag.

CARACO, T. 1980. Foraging time allocation in a stochastic environment. *Ecol.* 61: 119–28.

CARPENTER, S. R., J. F. KITCHELL, AND J. R. HODGSON. 1985. Cascading trophic interactions and lake productivity. *BioScience* 35:634–39.

CHAPIN, F. S., III, A. J. BLOOM, C. B. FIELD, AND R. H. WARING. 1987. Plant responses to multiple environmental factors. *BioScience* 37:49–57.

CHESSON, P., AND N. HUNTLEY. 1997. The roles of harsh and fluctuating conditions in the dynamics of ecological communities. *Am. Nat.* 150:519–53.

CLEMENTS, F. E. 1916. *Plant succession: Analysis of the development of vegetation.* Washington, D.C.: Carnegie Institute of Washington Publication, no. 242.

COHEN, J. E. 1989. Food webs and community structure. In *Perspectives in ecological theory*, ed. J. Roughgarden, R. M. May, and. S. A. Levin. Princeton: Princeton Univ. Press.

COHEN, J. E., AND F. BRIAND. 1984. Trophic links of community food webs. Proceedings of the National Academy of Sciences (USA) 81:105–4109.

COHEN, J. E., S. L. PIMM, P. YODZIS, AND J. SALDANA. 1993. Body sizes of animal predators and animal prey in food webs. *Journal of Animal Ecology.* 62:67–78.

COHEN, Y., T. L. VINCENT, AND J. S. BROWN. 1999. A G-function approach to fitness minima, fitness maxima, evolutionarily stable strategies, and adaptive landscapes. *Evolutionary Ecology Research.* 1:923–42.

CONNELL, J. H. 1961. The influence of interspecific competition and other factors on the distribution of the barnacle *Chthamalus stellatus. Ecol.* 42:710–23.

CONNELL, J. H., AND R. O. SLATYER. 1977. Mechanisms of succession in natural communities and their role in community stability and organization. *Am. Nat.* 111: 1119–44.

COUSINS, S. H. 1990. Countable ecosystems deriving from a new food web entity. *Oikos* 57:270–75.

COWLES, H. C. 1899. The ecological relations of the vegetation on the sand dunes of Lake Michigan. *Botanical Gazette* 27:95–117.

DAMUTH, J. 1981. Population density and body size in mammals. *Nature* 290:699–700.

DARWIN, C. 1859. *The origin of species by means of natural selection.* London: John Murray.

DAVIS, M. B. 1986. Climatic instability, time lags, and community disequilibrium. In *Community Ecology,* ed. J. Diamond and T. J. Case: 269–84. New York: Harper & Row.

DAYTON, P. K. 1971. Competition, disturbance and community organization: The provision and subsequent utilization of space in a rocky intertidal community. *Ecol. Monog.* 41:351–89.

DENSLOW, J. S. 1980. Patterns of plant species diversity during succession under different disturbance regimes. *Oecologia* 46:18–21.

DiBACCO, C., AND L. A. LEVIN. 2000. Development and application of elemental fingerprinting to track the dispersal of marine invertebrate larvae. *Limnology and Oceanography,* 45:871–80.

DOBZHANSKY, T. 1937. *Genetics and the origin of species.* New York: Columbia Univ. Press.

———. 1973. Nothing in biology makes sense except in the light of evolution. *American Biology Teacher* 35:125–29.

DOWNTON, M. W., AND K. A. MILLER. 1998. Relationships between Alaskan salmon catch and North Pacific climate on interannual and interdecadal time scales. *Canadian Journal of Fisheries and Aquatic Sciences* 55:2255–65.

DWYER, G., AND J. S. ELKINTON. 1995. Host dispersal and the spatial spread of insect pathogens. *Ecol.* 76:1262–75.

EHRLICH, P. R., R. R. WHITE, M. C. SINGER, S. W. McKECHNIE, AND L. E. GILBERT. 1975. Checkerspot butterflies: A historical perspective. *Science* 188:221–28.

ELKINTON, J. S., W. H. HEALY, J. P. BUONACCORSI, G. H. BOETTNER, A. M. HAZZARD, H. R. SMITH, AND A. M. LIEBHOLD. 1996. Interactions among gypsy moths, white-footed mice, and acorns. *Ecol.* 77:2332–42.

ELLNER, S., AND P. TURCHIN. 1995. Chaos in a noisy world: New methods and evidence from time-series analysis. *Am. Nat.* 145:343–75.

ELSER, J. J., M. M. ELSER, N. A. MACKAY, AND S. R. CARPENTER. 1988. Zooplankton-mediated transitions between N- and P-limited algal growth. *Limnol. Oceanog.* 33:1–14.

ELTON, C. 1935. Review of Theorie Analytique. *J. Anim. Ecol.* 4:148–50.

EMLEN, J. M. 1966. The role of time and energy in food preference. *Am. Nat.* 100: 611–17.

FAUTH, J. E., J. BERNARDO, M. CAMARA, W. J. RESTARITS, JR., J. VAN BUSKIRK, AND S. A. McCOLLUM. 1996. Simplifying the jargon of community ecology: A conceptual approach. *Am. Nat.* 147:282–86.

FISHER, R. A. 1930. *A genetical theory of natural selection.* Oxford: Clarendon.

FRETWELL, S. 1972. *Populations in a seasonal environment.* Princeton: Princeton Univ. Press.

FUHRMAN, J. A. 1999. Marine viruses and their biogeochemical and ecological effects. *Nature* 399:541–48.

GASTON, K. J. 1999. Implications of interspecific and intraspecific abundance-occupancy relationships. *Oikos* 86:195–207.

GASTON, K. J., AND T. M. BLACKBURN. 1999. A critique for macroecology. *Oikos* 84:353–68.

GASTON, K. J. AND J. H. LAWTON. 1988. Patterns in body size population dynamics and regional distribution of bracken herbivores. *Am. Nat.* 132:662–80.

GAUSE, G. F. 1934. *The struggle for existence.* Baltimore: Williams & Wilkins.

GILLIAM, J. F., AND D. F. FRASER. 1987. Habitat selection under predation hazard: A test of a model with foraging minnows. *Ecol.* 68:1856–62.

GLEASON, H. A. 1926. The individualistic concept of the plant association. *Torrey Botanical Club Bulletin* 53:7–26.

GLEESON, S. K., AND D. TILMAN. 1992. Plant allocation and the multiple limitation hypothesis. *Am. Nat.* 139:1322–43.

GLYNN, P. W. 1988. El Niño–southern oscillation 1982–83: Nearshore population, community, and ecosystem responses. *Annual Review of Ecology and Systematics* 19:309–45.

GOTELLI, N. J. 1991. Metapopulation models: The rescue effect, the propagule rain, and the core-satellite hypothesis. *Am. Nat.* 138:768–76.

GOTELLI, N. J., AND G. R. GRAVES. 1996. *Null models in ecology.* Washington, D.C.: Smithsonian Institution Press.

GRACE, J. B., AND B. H. PUGESEK. 1997. A structural equation model of plant species richness and its application to a coastal wetland. *Am. Nat.* 149:436–60.

———. 1998. On the use of path analysis and related procedures for the investigation of ecological problems. *Am. Nat.* 152:151–59.

GRINNELL, J. 1917. The niche relationships of the California thrasher. *Auk* 21:364–82.

GROSHOLZ, E. D. 1996. Contrasting rates of spread for introduced species in terrestrial and marine systems. *Ecol.* 77:1680–86.

HAECKEL, E. 1869. Entwicklungsgang und Aufgaben der Zoologie. *Jenaische Zeitschrift* 5:353.

HAIRSTON, N. G., JR., AND N. G. HAIRSTON, SR. 1997. Does food web complexity eliminate trophic-level dynamics? *Am. Nat.* 149:1001–7.

HAIRSTON, N. G., F. E. SMITH, AND L. B. SLOBODKIN. 1960. Community structure, population control, and competition. *Am. Nat.* 94:421–25.

HANSKI, I., AND M. GILPIN, EDS. 1997. Metapopulation biology: Ecology, genetics and evolution. New York: Academic Press.

HART, D. D., S. L. KOHLER, AND R. G. CARLTON. 1991. Harvesting of benthic algae by territorial grazers the potential for prudent predation. *Oikos* 60:329–35.

HARVEY, P. H., AND R. M. ZAMMUTO. 1985. Patterns of mortality and age at first reproduction in natural populations of mammals. *Nature* 315:319–20.

HASTINGS, A., C. L. HOM, S. ELLNER, P. TURCHIN, AND H. C. J. GODFRAY. 1993. Chaos in ecology: Is Mother Nature a strange attractor? *Ann. Rev. Ecol. Syst.* 24:1–33.

HEINSELMAN, M. L. 1973. Fire in the virgin forests of the Boundary Waters Canoe Area, Minnesota. *Quaternary Research* 3:329–82.

HIGASHI, M., T. P. BURNS, AND B. C. PATTEN. 1993. Network trophic dynamics: The tempo of energy movement and availability in ecosystems. *Ecological Modelling* 66:43–64.

HILS, M. H., AND J. L. VANKAT. 1982. Species removals from a 1ˢᵗ year old field community. *Ecol.* 63:705–11.

HOLMES, E. E. 1993. Are diffusion models too simple? A comparison with telegraph models of invasion. Am. Nat. 142:779–95.

HOLT, R. D. 1996. Adaptive evolution in source-sink environments: Direct and indirect effects of density-dependence on niche evolution. Oikos 75: 182–92.

HUFFAKER, C. B. 1958. Experimental studies on predation: Dispersion factors and predator-prey oscillations. Hilgardia 27:343–83.

HUNTER, M. D., AND P. W. PRICE. 1992. Playing chutes and ladders: Bottom-up and top-down forces in natural communities. Ecol. 73:724–32.

HUTCHINSON, G. E. 1957. Concluding remarks. Cold Spring Harbor Symposium on Quantitative Biology 22:415–27.

———. 1959. Homage to Santa Rosalia or why are there so many kinds of animals? Am. Nat. 93:145–59.

———. 1961. The paradox of the plankton. Am. Nat. 95:137–45.

JANZEN, D. H. 1967. Why mountain passes are higher in the tropics. Am. Nat. 101: 233–49.

JONES, C. G., J. H. LAWTON, AND M. SHACHAK. 1994. Organisms as ecosystem engineers. Oikos 69:373–86.

JONES, C. G., R. S. OSTFELD, M. P. RICHARD, E. M. SCHAUBER, AND J. O. WOLFF. 1998. Chain reactions linking acorns to gypsy moth outbreaks and lyme disease risk. Science 279:1023–26.

JONGMAN, R. H. G., C. J. F. TER BRAAK, AND O. F. R. VAN TONGEREN, EDS. 1995. Data analysis in community and landscape ecology. Cambridge: Cambridge Univ. Press.

KAREIVA, P. 1987. Habitat fragmentation and the stability of predator-prey interactions. Nature 326:388–90.

KERMACK, W. O., AND A. G. McKENDRICK. 1927. A contribution to the mathematical theory of epidemics. Proceedings of the Royal Society of London A 115:700–21.

KOTLER, B. P., J. S. BROWN, R. H. SLOTWO, W. L. GOODFRIEND, AND M. STRAUSS. 1993. The influence of snakes on the foraging behavior of gerbils. Oikos 67:309–16.

LACK, D. 1954. The natural regulation of animal numbers. Oxford: Oxford Univ. Press.

LANE, P., AND R. LEVINS. 1977. Dynamics of aquatic systems. 2. The effects of nutrient enrichment on model plankton communities. Limnol. Oceanog. 22:454–71.

LASKA, M. S., AND J. T. WOOTTON. 1998. Theoretical concepts and empirical approaches to measuring interaction strength. Ecol. 79:461–76.

LAWLOR, L. R. 1979. Direct and indirect effects of n-species competition. Oecol. 43: 355–64.

LEGENDRE, P., R. GALZIN, AND M. L. HARMELIN-VIVIEN. 1997. Relating behavior to habitat: Solution to the fourth-corner problem. Ecol. 78:547–62.

LEGENDRE, P., AND L. LEGENDRE. 1998. Numerical ecology. 2d English ed. Amsterdam: Elsevier.

LEIBOLD, M. A. 1989. Resource edibility and the effects of predators and productivity on the outcome of trophic interactions. Am. Nat. 134:922–49.

———. 1995. The niche concept revisited: Mechanistic models and community context. Ecol. 76:1371–82.

———. 1996. A graphical model of the keystone predators in food webs: Trophic regulation of abundance, incidence, and diversity patterns in nature. Am. Nat. 147: 784–812.

———. 1999. Biodiversity and nutrient enrichment in pond plankton communities. *Evolutionary Ecology Research* 1:73–95.

Leibold, M. A., J. M. Chase, J. Shurin, and A. Downing. 1997. Species turnover and the regulation of trophic structure. *Ann. Rev. Ecol. Syst.* 28:467–97.

Leibold, M. A., and A. J. Tessier. 1991. Contrasting patterns of body size for *Daphnia* species that segregate by habitat. *Oecol.* 86:342–48.

———. 1998. Experimental compromise and mechanistic approaches to the evolutionary ecology of interacting *Daphnia* species. In *Experimental ecology: Issues and perspectives*, ed. W. J. Resetarits, Jr. and J. Bernardo: 96–112. Oxford: Oxford Univ. Press.

Levine, S. H. 1976. Competitive interactions in ecosystems. *Am. Nat.* 110:903–10.

Levins, R. 1969. Some demographic and genetic consequences of environmental heterogeneity on biological control. *Bulletin of the Entomological Society* 15:237–40.

———. 1975. Evolution in communities near equilibrium. In *Ecology and evolution of communities*, ed. M. L. Cody and J. M. Diamond: 16–50. Cambridge: Harvard Univ. Press.

———. 1979. Coexistence in a variable environment. *Am. Nat.* 114:765–83.

Levitan, C. 1987. Formal stability analysis of a planktonic freshwater community. In *Predation: Direct and indirect impacts on aquatic communities*, ed. W. C. Kerfoot and A. Sih: 71–100. Hanover, N.H.: Univ. Press of New England.

Lewis, M. A. 1997. Variablity, patchiness, and jump dispersal in the spread of an invading population. In *Spatial ecology: The role of space in population dynamics and interspecific interactions*, ed. D. Tilman and P. Kareiva: 46–74. Princeton: Princeton Univ. Press.

Lindeman, R. L. 1942. The trophic-dynamic aspect of ecology. *Ecol.* 23:399–418.

Loreau, M., and N. Mouquet. 1999. Immigration and the maintenance of local species diversity. *Am. Nat.* 154: 427–40.

Lotka, A. J. 1925. *Elements of physical biology*. Baltimore: Williams & Wilkins.

Lubchenco, J. 1983. *Littorina* and *Fucus*: Effects of herbivores, substratum heterogeneity and plant escapes during succession. *Ecol.* 64:1116–23.

Lubina, J. A., and S. A. Levin. 1988. The spread of a reinvading species range expansion in the California sea otter. *Am. Nat.* 131:526–43.

MacArthur, R. H. 1955. Fluctuations of animal populations, and a measure of community stability. *Ecol.* 36:533–36.

MacArthur, R. H., and R. Levins. 1967. The limiting similarity, convergence and divergence of coexisting species. *Am. Nat.* 101:377–85.

MacArthur, R. H., and E. R. Pianka. 1966. On optimal use of a patchy environment. *Am. Nat.* 100:603–9.

MacArthur, R. H., and E. O. Wilson. 1969. *The theory of island biogeography.* Princeton: Princeton Univ. Press.

Mantua, N. J., S. R. Hare, Y. Zhang, J. M. Wallace, and R. C. Francis. 1997. A Pacific interdecadal climate oscillation with impacts on salmon production. *Bulletin of the American Meteorological Society* 78:1069–80.

Martin, T. H., L. B. Crowder, C. F. Dumas, and J. M. Burkholder. 1992. Indirect effects of fish on macrophytes in Bays Mountain Lake: Evidence for a littoral trophic cascade. *Oecol.* 89:476–81.

MARTINEZ, N. D. 1994. Scale-dependent constraints on food web structure. Am. Nat. 144:935–53.

MAY, R., AND G. OSTER. 1976. Bifurcations and dynamic complexity in simple ecological models. Am. Nat. 110:573–99.

MAYR, E. 1942. Systematics and the origin of species from the viewpoint of a zoologist. New York: Columbia Univ. Press.

McCANN, K., A. HASTINGS, AND G. HUXEL. 1998. Weak trophic interactions and the balance of nature. Nature 395:794–98.

McCAULEY, E., AND F. BRIAND. 1979. Zooplankton grazing and phytoplankton species richness: Field tests of the predation hypothesis. Limnol. Oceang. 24: 243–52.

McGOWAN, J. A., D. R. CAYAN, AND L. M. DORMAN. 1988. Climate–ocean variability and ecosystem response in the northeast. Science 281:210–17.

McPEEK, M. A. 1998. The consequences of changing the top predator in a food web: A comparative experimental approach. Ecol. Monog. 68:1–23.

MENGE, B. A. 1995. Indirect effects in marine rocky intertidal interaction webs: Patterns and importance. Ecol. Monog. 65:21–74.

MENGE, B. A., AND T. M. FARRELL. 1989. Community structure and interaction webs in shallow marine hard-bottom communities: Tests of an environmental stress model. Advances in Ecological Research 19:189–262.

MILLER, T. E. 1994. Direct and indirect species interactions in an early old-field plant community. Am. Nat. 143:1007–25.

MORIN, P. J. 1995. Functional redundancy, non-additive interactions, and supply-side dynamics in experimental pond communities. Ecol. 76:133–49.

MORRIS, W. F., AND D. M. WOOD. 1989. The role of lupine in succession on Mt. St. Helens, Washington, USA: Facilitation or inhibition? Ecol. 70:697–703.

NEILL, W. E. 1974. The community matrix and interdependence of the competition coefficients. Am. Nat. 108:399–408.

NICHOLSON, A. J., AND V. A. BAILEY. 1935. The balance of animal populations. Proceedings of the Zoological Society of London:551–98.

NISBET, R. M., W. S. C. GURNEY, W. W. MURDOCH, AND E. McCAULEY. 1989. Structured population models: A tool for linking effects at the individual and population levels. Biological Journal of the Linnean Society 37:79–99.

ODUM, E. P. 1953. Fundamentals of ecology. Philadelphia: Saunders.

OKSANEN, L., S. D. FRETWELL, J. ARRUDA, AND P. NIEMELA. 1981. Exploitation ecosystems in gradients of primary productivity. Am. Nat. 118:240–61.

OKUBO, A. 1980. Diffusion and ecological problems: Mathematical models. Lecture notes in Biomathematics 10. Heidelberg, Germany: Springer-Verlag.

ORIANS, G. H. 1962. Natural selection and ecological theory. Am. Nat. 96:257–64.

OSENBERG, C. W., AND G. G. MITTELBACH. 1996. The relative importance of resource limitation and predator limitation in food chains. In Food webs: Integration of pattern and dynamics, ed. G. A. Polis and K. O. Winemiller: 134–48. New York: Chapman & Hall.

PACKER, C., D. SCHEEL, AND A. E. PUSEY. 1990. Why lions form groups: Food is not enough. Am. Nat. 136:1–19.

PAHL-WOSTL, C. 1995. The dynamic nature of ecosystems: Chaos and order entwined. Chichester, N.Y.: Wiley.

PAINE, R. T. 1966. Food web complexity and species diversity. *Am. Nat.* 100:65–75.

———. 1980. Food webs: Linkage, interaction strength and community infrastructure. *J. Anim. Ecol.* 49:667–85.

———. 1988. Food webs: Road maps of interactions or grist for theoretical development? *Ecol.* 69:1648–54.

———. 1992. Food web analysis through field measurement of per capita interaction strength. *Nature* 355:73–75.

PAINE, R. T., AND S. A. LEVIN. 1981. Intertidal landscapes: Disturbance and the dynamics of pattern. *Ecol. Monog.* 51:145–78.

PATTERSON, B. D., AND W. ATMAR. 1986. Nested subsets and the structure of insular mammalian faunas and archipelagos. *Biol. J. Linnean Society* 28:65–82.

PETERS, R. H. 1983. *The ecological implications of body size.* Cambridge: Cambridge Univ. Press.

PETRAITIS, P. S., A. E. DUNHAM, AND P. H. NIEWIAROWSKI. 1996. Inferring multiple causality: The limitations of path analysis. *Functional Ecology* 10:421–31.

PFISTER, C. A. 1996. The role and importance of recruitment variability to a guild of tide pool fishes. *Ecology* 77:1928–41.

———. 1998. Patterns of variance in stage-structured populations: Evolutionary predictions and ecological implications. *Proceedings of the National Academy of Science (USA)* 95:213–18.

PHILLIPS, O. M. 1974. The equilibrium and stability of simple marine systems. II Herbivores. *Archivs für Hydrobiologie* 73:310–33.

PICKETT, S. T. A., AND P. S. WHITE, EDS. 1985. *The ecology of natural disturbance and patch dynamics.* New York: Academic Press.

PIELOU, E. C. 1984. *The interpretation of ecological data: A primer on classification and ordination.* New York: Wiley.

PIMM, S. L. 1979. The structure of food webs. *Theor. Pop. Biol.* 16:144–58.

———. 1982. *Food webs.* London: Chapman & Hall.

PIMM, S. L., AND J. H. LAWTON. 1978. On feeding at more than one trophic level. *Nature* 275:542–44.

PIONTKOVSKI, S. A., R. WILLIAMS, AND T. A. MELNIK. 1995. Spatial heterogeneity, biomass and size structure of plankton of the Indian Ocean: Some general trends. *Marine Ecology—Progress Series.* 117:219–27.

PLATT, W. J. 1975. The colonization and formation of equilibrium plant species associations on badger disturbances in a tall-grass prairie. *Ecol. Monog.* 45:285–305.

POLIS, G. A. 1991. Complex trophic interactions in deserts: An empirical critique of food web theory. *Am. Nat.* 138:123–55.

POLIS, G. A., AND R. D. HOLT. 1992. Intraguild predation the dynamics of complex trophic interactions. *Trends in Ecology & Evolution* 7:151–54.

POLIS, G. A., AND S. D. HURD. 1996. Allochthounous input across habitats, subsidized consumers, and apparent trophic cascades: Examples from the ocean–land interface. In *Food webs: Integration of pattern and dynamics,* ed. G. A. Polis and K. O. Winemiller: 275–85. New York: Chapman & Hall.

Polis, G. A., and K. O. Winemiller, eds. 1996. *Food webs: Integration of patterns and dynamics*. New York: Chapman & Hall.

Power, M. E. 1992. Top-down and bottom-up forces in food webs: Do plants have primacy? *Ecology* 73:733–46.

Preston, F. W. 1962. The canonical distribution of commonness and rarity. *Ecol.* 43:185–215, 410–32.

Price, P. W. 1975. *Insect ecology*. New York: Wiley Interscience.

Pugesek, B. H., and J. B. Grace. 1998. On the utility of path modelling for ecological and evolutionary studies. *Funct. Ecol.* 12:853–56.

Real, L. A. 1980. Fitness, uncertainty, and the role of diversification in evolution and behavior. *Am. Nat.* 115:623–38.

Resetarits, W. J., and J. Bernardo, eds. 1998. *Issues and perspectives in experimental ecology*. New York: Oxford Univ. Press.

Reynolds, H. L., and S. W. Pacala. 1993. An analytical treatment of root-to-shoot ratio and plant competition for soil nutrients and light. *Am. Nat.* 141:51–70.

Root, R. B. 1973. Organization of a plant-arthropod association in simple and diverse habitats: The fauna of collards (*Brassica oleracea*). *Ecol. Monog.* 43:95–124.

Rosenzweig, M. L. 1981. A theory of habitat selection. *Ecol.* 62:327–35.

Roughgarden, J., S. Gaines, and H. Possingham. 1988. Recruitment dynamics in complex life cycles. *Science* 241:1460–66.

Roy, K. D. Jablonski, and J. W. Valentine. 1994. Eastern Pacific molluscan provinces and latitudinal diversity gradients: No evidence for "Rapoport's rule." Proceedings of the National Academy of Sciences (USA) 91:8871–74.

Schaffer, W. M. 1981. Ecological abstraction: The consequences of reduced dimensionality in ecological models. *Ecol. Monog.* 51:383–401.

Schmitz, O. J. 1997. Press perturbations and the predictability of ecological interactions in a food web. *Ecol.* 78:55–69.

Schoener, T. W. 1989. The ecological niche. In *Ecological concepts: The contribution of ecology to an understanding of the natural world*, ed. J. M Cherrett: 79–447. Oxford: Blackwell Scientific.

Shelford, V. E. 1913. *Animal communities in temperate America*. Chicago: Univ. of Chicago Press.

Sih, A. 1992. Prey uncertainty and the balancing of antipredator and feeding needs. *Am. Nat.* 139:1052–69.

Sih, A., J. W. Petranka, and L. B. Kats. 1988. The dynamics of prey refuge use: A model and tests with sunfish and salamander larvae. *Am. Nat.* 132:463–83.

Simpson, G. G. 1944. *Tempo and mode in evolution*. New York: Columbia Univ. Press.

Skellam, J. G. 1951. Random dispersal in theoretical populations. *Biometrika* 38: 196–218.

Smith, F. A., J. H. Brown, and T. J. Valone. 1997. Path analysis: A critical evaluation using long-term experimental data. *Am. Nat.* 149:29–42.

Smith, V. H. 1983. Low nitrogen to phosphorus ratios favor dominance by blue-green algae in lake phytoplankton. *Science* 221:669–72.

Sousa, W. P. 1979. Disturbance in marine intertidal boulder fields: The nonequilibrium maintenance of species diversity. *Ecol.* 60:1225–39.

SPILLER, D. A., AND T. W. SCHOENER. 1990. A terrestrial field experiment showing the impact of eliminating top predators on foliage damage. *Nature* 347:469–72.

STEPHENS, D. W., AND J. R. KREBS. 1986. *Foraging theory*. Princeton: Princeton Univ. Press.

STERNER, R. W. 1994. Elemental stoichiometry of species in ecosystems. In *Linking species and ecosystems*, ed. C. G. Jones and J. H. Lawton: 240–52. New York: Chapman & Hall.

STEVENS, G. C. 1989. The latitudinal gradient in geographical range: How so many species coexist in the tropics. *Am. Nat.* 133, 240–56.

STRONG, D. R. 1986. Density vagueness: Abiding the variance in the demography of real populations. In *Community ecology*, ed. J. Diamond and T. J. Case: 257–68. New York: Harper & Row.

STRONG, D. R., JR., D. SIMBERLOFF, L. G. ABELE, AND A. B. THISTLE, EDS. 1984. *Ecological communities: Conceptual issues and the evidence*. Princeton: Princeton Univ. Press.

SWEARER, S. E., J. E. CASELLE, D. W. LEA, AND R. R. WARNER. 1999. Larval retention and recruitment in an island population of a coral-reef fish. *Nature* 402: 799–802.

TANSLEY, A. G. 1923. *Practical plant ecology*. New York: Dodd, Mead & Co.

TANSLEY, A. G., AND R. S. ADAMSON. 1925. Studies of the vegetation of the English chalk. III. The chalk grasslands of the Hampshire–Sussex border. *Journal of Ecology* 5:173–79.

TILMAN, D. 1982. *Resource competition and community structure*. Princeton: Princeton Univ. Press.

———. 1988. *Plant strategies and the dynamics and structure of plant communities*. Princeton: Princeton Univ. Press.

———. 1994. Competition and biodiversity in spatially structured habitats. *Ecol.* 75:2–16.

———. 1997. Community invasibility, recruitment limitation and grassland biodiversity. *Ecol.* 78:81–92.

UNDERWOOD, A. J. 1978. A refutation of critical tidal levels as determinants of the structure of intertidal communities on British shores. *Journal of Experimental Marine Biology & Ecology* 33:261–76.

VANNOTE, R. L., G. W. MINSHALL, K. W. CUMMINS, J. R. SEDELL, AND C. E. CUSHING. 1980. The river continuum concept. *Canadian Journal of Fisheries and Aquatic Science* 37:130–37.

VAN VALEN, L. M. 1973. Body size and the numbers of plants and animals. *Evolution* 27:27–35.

VARLEY, G. C. 1947. The natural control of population balance in the kanpweed gallfly (*Urophora jaceana*). *J. Anim. Ecol.* 16:139–87.

VEIT, R. R., AND M. A. LEWIS. 1996. Dispersal, population growth and the Allee effect: Dynamics of the house finch invasion of North America. *Am. Nat.* 148:255–74.

VOLTERRA, V. 1931. Variations and fluctuations of the numbers of individuals in animal species living together. Translated in *Animal ecology*, ed. R. N. Chapman. New York: McGraw-Hill.

Von Liebig, J. 1855. *Die Grundsatze der Agriculturchemie.* Braunschweig, Germany: Vieweig.

Ward, J. V., and J. A. Stanford. 1983. The intermediate-disturbance hypothesis: An explanation for biotic diversity patterns in lotic ecosystems. In *Dynamics of lotic ecosystems,* ed. T. D. Fontaine and S. M. Bartell: 347–56. Ann Arbor, Mich.: Ann Arbor Science Publishers.

Werner, E. E., and B. Anholt. 1996. Predator-induced behavioral indirect effects: Consequences to competitive interactions in anuran larvae. *Ecol.* 77:157–69.

Werner, E. E., and J. F. Gilliam. 1984. The ontogenetic niche and species interactions in size-structured populations. *Ann. Rev. Ecol. Syst.* 15:393–425.

Whitaker, R. H. 1956. Vegetation of the Great Smoky Mountains. *Ecol. Monog.* 23:41–78.

Wilbur, H. M. 1987. Regulation of structure in complex experimental temporary pond communities. *Ecol.* 68:1437–52.

Wilbur, H. M., and J. E. Fauth. 1990. Experimental aquatic food webs: Interactions between two predators and two prey. *Am. Nat.* 135:176–204.

Wilson, E. O. 1988. The current state of biodiversity. In *Biodiversity,* ed. E. O. Wilson: 1–18. Washington, D.C.: National Academy Press.

Wootton, J. T. 1987. The effects of body mass, phylogeny, habitat and trophic level on mammalian age at first reproduction. *Evol.* 41:732–49.

———. 1992. Indirect effects, prey susceptibility, and habitat selection: Impacts of birds on limpets and algae. *Ecol.* 73:981–91.

———. 1993. Indirect effects and habitat use in an intertidal community: Interaction chains and interaction modifications. *Am. Nat.* 141:71–89.

———. 1994a. The nature and consequences of indirect effects in ecological communities. *Ann. Rev. Ecol. Syst.* 25:443–66.

———. 1994b. Predicting direct and indirect effects: An integrated approach using experiments and path analysis. *Ecol.* 75:151–65.

Wootton, J. T., and M. E. Power. 1993. Productivity, consumers and the structure of a river food chain. *Proceedings of the National Academy of Science* 90:1384–87.

Zaret, T. 1980. *Predation and freshwater communities.* New Haven, Conn.: Yale Univ. Press.

ANIMAL ECOLOGY

CHAPTER I

INTRODUCTION

" Faunists, as you observe, are too apt to acquiesce in bare descriptions and a few synonyms ; the reason for this is plain, because all that may be done at home in a man's study, but the investigation of the life and conversation of animals is a concern of much more trouble and difficulty, and is not to be attained but by the active and inquisitive, and by those that reside much in the country."—GILBERT WHITE, 1771.

1. ECOLOGY is a new name for a very old subject. It simply means scientific natural history. To a great many zoologists the word "natural history" brings up a rather clear vision of parties of naturalists going forth on excursion, prepared to swoop down on any rarity which will serve to swell the local list of species. It is a fact that natural history has fallen into disrepute among zoologists, at any rate in England, and since it is a very serious matter that a third of the whole subject of zoology should be neglected by scientists, we may ask for reasons. The discoveries of Charles Darwin in the middle of the nineteenth century gave a tremendous impetus to the study of species and the classification of animals. Although Linnæus had laid the foundation of this work many years before, it was found that previous descriptions of species were far too rough and ready, and that a revision and reorganisation of the whole subject was necessary. It was further realised that many of the brilliant observations of the older naturalists were rendered practically useless through the insufficient identification of the animals upon which they had worked. Half the zoological world thereupon drifted into museums

and spent the next fifty years doing the work of description and classification which was to lay the foundations for the scientific ecology of the twentieth century. The rest of the zoologists retired into laboratories, and there occupied their time with detailed work upon the morphology and physiology of animals. It was an age of studying whole problems on many animals, rather than the whole biology of any one animal. The morphologist does not require the identification of his specimens below orders or families (in extreme cases) genera. The physiologist takes the nearest convenient animal, generally a parasite or a pet of man, and works out his problems on them. The point is that most morphology and physiology could be done without knowing the exact name of the animal which was being studied, while ecological work could not. Hence the temporary dying down of scientific work on animal ecology.

2. Meanwhile a vast number of local natural history societies burst into bloom all over Britain, and these bent their energies towards collecting and storing up in museums the local animals and plants. This work was of immense value, as it provided the material for classifying animals properly. But as time went on, and the groundwork of systematics was covered and consolidated, the collecting instinct went through the various stages which turn a practical and useful activity into a mania. At the present day, local natural history societies, however much pleasure they may give to their members, usually perform no scientific function, and in many cases the records which are made are of less value than the paper upon which they are written. Miall commented on this fact as long ago as 1897 when he said: "Natural history . . . is encumbered by multitudes of facts which are recorded only because they are easy to record."[79]* Such is the history of these societies. Like the bamboo, they burst into flower, produced enormous masses of seed, and then died with the effort. But however this may be, it is necessary for zoologists to realise that the work of the last fifty years has made field work on animals a practical possibility.

* The small numbers refer to the bibliography at the end of this book.

on an animal unless you knew its name. Scientific ecology was first started some thirty years ago by botanists, who finished their classification sooner than the zoologists, because there are fewer species of plants than of animals, and because plants do not rush away when you try to collect them. Animal ecologists have followed the lead of plant ecologists and copied most of their methods, without inventing many new ones of their own. It is one of the objects of this book to show that zoologists require quite special methods of their own in order to cope properly with the problems which face them in animal ecology.

3. When we take a broad historical view, it becomes evident that men have studied animals in their natural surroundings for thousands of years—ever since the first men started to catch animals for food and clothing; that the subject was developed into a science by the brilliant naturalists before and at the time of Charles Darwin; and that the discoveries of Darwin, himself a magnificent field naturalist, had the remark-able effect of sending the whole zoological world flocking indoors, where they remained hard at work for fifty years or more, and whence they are now beginning to put forth cautious heads again into the open air. But the open air feels very cold, and it has become such a normal proceeding for a zoologist to take up either a morphological or physiological problem that he finds it rather a disconcerting and disturbing experience to go out of doors and study animals in their natural conditions. This is not surprising when we consider that he has never had any opportunity of becoming trained in such work. In spite of this, the work badly needs doing; the fascination of it lies in the fact that there are such a number of interesting problems to be found, so many to choose from, and requiring so much energy and resource to solve. Adams says : " Here, then, is a resource, at present largely unworked by many biologists, where a wealth of ideas and explanations lies strewn over the surface and only need to be picked up in order to be utilised by those acquainted with this method of interpretation," [1b] while Tansley, speaking of plant ecology, says : " Every genuine worker in science is an explorer, who is continually

meeting fresh things and fresh situations, to which he has to adapt his material and mental equipment. This is conspicuously true of our subject, and is one of the greatest attractions of ecology to the student who is at once eager, imaginative, and determined. To the lover of prescribed routine methods with the certainty of 'safe' results the study of ecology is not to be recommended." [15a]

CHAPTER II

THE DISTRIBUTION OF ANIMAL COMMUNITIES

Each habitat (1) has living in it a characteristic community of animals; (2) these can be classified in various ways and (3) their great variety and richness is due to the comparative specialisation of most species of animals. (4) It is convenient to study the zonation of such communities along the various big gradients in environmental conditions, such as that from the poles to the equator, which (5) shows the dominating influence of plants upon the distribution of animals, in forming special local conditions and (6) by producing sharp boundaries to the habitats so that (7) animal communities are more sharply separated than they would otherwise be. This is clearly shown by (8) the vertical zones of communities on a mountain-side, which also illustrate the principle that (9) the members of each community can be divided into those "exclusive" to and (10) into those "characteristic" of it, while the remaining species, which form the bulk of the community, occur in more than one association. (11) Other vertical gradients are that of light in the sea and (12) that of salts in water. (13) Each large zone can be subdivided into smaller gradients of habitat, *e.g.* water-content of the soil, and (14) these again into still smaller ones, until we reach single species of animals, which in turn can be shown to contain gradients in internal conditions supporting characteristic communities of parasites. (15) In such ways the differences between communities can be classified and studied as a preliminary to studying the fundamental resemblances amongst them.

1. ONE of the first things with which an ecologist has to deal is the fact that each different kind of habitat contains a characteristic set of animals. We call these animal associations, or better, animal communities, for we shall see later on that they are not mere assemblages of species living together, but form closely-knit communities or societies comparable to our own. Up to the present time animal ecologists have been very largely occupied with a general description and classification of the various animal habitats and of the fauna living in them. Preliminary biological surveys have been undertaken in most civilised countries except England and China, where animal ecology lags behind in a peculiar way. In particular

we might mention the work of the American Bureau of
Biological Survey (which was first started in order to study
problems raised by the introduction of the English sparrow
into the United States), and of other institutions in that country,
and similar surveys undertaken under the initiative of the late
Dr. Annandale in India. It is clearly necessary to have a list
of the animals in different habitats before one can proceed to
study the more intricate problems of animal communities.
We shall return to the question of biological surveys in the
chapter on "Methods."

2. Various schemes have been proposed for the classification of animal communities, some very useful and others completely absurd. Since, however, no one adopts the latter,
they merely serve as healthy examples of what to avoid, namely,
the making of too many definitions and the inventing of a host
of unnecessary technical terms. It should always be remembered that the professional ecologist has to rely, and always
will have to rely, for a great many of his data, upon the observations of men like fishermen, gamekeepers, local naturalists,
and, in fact, all manner of people who are not professional
scientists at all. The life, habits, and distribution of animals
are often such difficult things to ascertain and so variable from
time to time, that it will always be absolutely essential to use
the unique knowledge of men who have been studying animals
in one place for a good many years. It is a comparatively
simple matter to make a preliminary biological survey and
accumulate lists of the animals in different communities.
This preliminary work requires, of course, great energy and
perseverance, and a skilled acquaintance with the ways of
animals ; but it is when one penetrates into the more intimate
problems of animal life, and attempts to construct the food-
cycles which will be discussed later on, that the immensity of
the task begins to appear and the difficulty of obtaining the
right class of data is discovered. It is therefore worth em-
phasising the vital importance of keeping in touch with all
practical men who spend much of their lives among wild
animals. To do this effectually it is desirable that ecology
should not be made to appear much more abstruse and difficult

than it really is, and that it should not be possible to say that "ecology consists in saying what every one knows in language that nobody can understand." The writer has learnt a far greater number of interesting and invaluable ecological facts about the social organisation of animals from gamekeepers and private naturalists, and from the writings of men like W. H. Hudson, than from trained zoologists. There is something to be said for the view of an anonymous writer in *Nature*, who wrote: "The notion that the truth can be sought in books is still widely prevalent and the present dearth of illiterate men constitutes a serious menace to the advancement of knowledge." Even if this is so, it is at the same time true that there is more ecology in the Old Testament or the plays of Shakespeare than in most of the zoological textbooks ever published !

All this being so, there seems to be no point in making elaborate and academic classifications of animal communities. After all, what is to be said for a scientist who calls the community of animals living in ponds the "tiphic asso- ciation," or refers to the art of gardening as "chronic hemerecology"?

3. It is important, however, to get some general idea of the variety and distribution of the different animal communities found in the world. The existence of such a rich variety of communities is to be attributed to two factors. In the first place, no one animal is sufficiently elastic in its organisation to withstand the wide range of environmental conditions which exist in the world, and secondly, nearly all animals tend during the course of evolution to become more or less specialised for life in a narrow range of environmental conditions, for by being so specialised they can be more efficient. This tendency towards specialisation is abundantly shown throughout the fossil record and is reflected in the numerous and varied animal communities of the present day. Primarily, there is specialisation to meet particular climatic and other physical and chemical factors, and secondarily, animals become adapted to a special set of biotic conditions—food, enemies, etc. The effect of the various environmental factors upon animals is a

subject which will be followed up separately in the next chapter.

4. One way of giving some idea of the range of different animal habitats, and of the communities living in them, is to take some of the big gradients in environment and show how the communities change as we pass from one end to the other. The biggest of these is the gradient in temperature and light intensity between the poles and the equator, which owes its existence to the globular shape of the world. At the one extreme there are the regions of polar ice-pack, with their peculiar animal communities living in continuous daylight during the summer and continuous darkness in the winter, and with corresponding abrupt seasonal changes in temperature. At the other end of the scale we have equatorial rain forest with a totally different set of animals adapted to life in a continuously hot climate, and in many cases in continuous semi-darkness, in the shade of the tropical trees. An animal like Bosman's Potto sees less light throughout the year than the Arctic fox. In between these extremes we have animal communities accustomed to a moderate amount of heat and light.

5. These examples serve to introduce a very important idea, namely, the effect of vegetation upon the habitats and distribution of animals.

Although a tropical rain forest partly owes its existence to the intense sunlight of the tropics, yet inside the forest it is quite dark, and it is clear that plants have the effect of translating one climate into another, and that an animal living in or under a plant community and dependent on it, may be living under totally different climatic conditions from those existing outside. Each plant association therefore carries with it, or rather in it, a special local "climate" which is peculiar to itself. Broadly speaking, plants have a blanketing effect, since they cut off rain, and radiant energy like light and heat. Their general effect is therefore to tone down the intensity of any natural climate. At the same time they reduce the amount of evaporation from the soil surface, and so make the air damper than it otherwise would be.

Looking at the matter very broadly, in the far north there

is very sparse vegetation, which does not always cover the ground at all completely. As we go south the vegetation becomes more dense and higher until we reach a zone with scattered trees. These are separated by wide intervals owing to the fact that the soil is too shallow (being frozen below a certain depth) to allow of sufficient root development except by extensive growth sideways. Then we find true forests, but still not very luxuriant. Finally, there are the immense rain forests of the equatorial belt. The gradation consists essentially first of a gradual filling up of the soil by roots, and then a covering up of the surface by vegetation.

6. There is a further important way in which plant communities affect animals. When we look at two plant communities growing next to one another, it is usually noticeable that the junction between the two is comparatively sharply marked. Examples of this are the zones of vegetation round the edge of a lake or up the side of a mountain. The reason for this sharp demarcation between plant communities is simple. Plants are usually competing for light, and if one plant in a community manages to outstrip the others in its growth it is able to cut off much of the light from them, and it then becomes *dominant*. This is the condition found in most temperate plant communities. Examples are the common heather (*Calluna*), or the beech trees in a wood, or the rushes (*Juncus*) in a marshy area. The process of competition is not always so simple as this, and there may be all manner of complicated factors affecting the relative growth of the competing species, but the final battle is usually for light. In some cases the winning species kills off other competitors, not by shading them, but by producing great quantities of dead leaves which swamp the smaller plants below, or by some other means. The main phenomenon of dominance remains the same. Now at the junctions of two plant communities there is also a battle for light going on, and it resolves itself mainly into a battle between the two respective dominants. Just as within the community, so between two different dominants, no compromise is possible in a battle for light. If one plant wins, it wins completely. Now every plant has a certain set

of optimum conditions for maximum growth, and as conditions depart from this optimum, growth becomes less efficient. Since each dominant has different optimum conditions, there is always a certain point in an environmental gradient where one dominant, and therefore one community, changes over fairly abruptly into another. There may be originally a regular and gradual gradient in, say, water-content of the soil, as from the edge of a lake up on to a dry moor; but the existence of dominance in plants causes this to be transformed into a series of sharply marked zones of vegetation, which to some extent mask the original gradient, and may even react on the surroundings so as to convert the conditions themselves into a step-like series.

7. It is clear, then, that because green plants feed by means of sunlight, the boundaries of their communities tend to be rather sharply defined ; and since we have seen that each plant community carries with it a special set of "climatic" conditions for the animals living in it, the rather sudden difference in conditions at the edges of the plant communities will be reflected in the animals. This means that the species of animals will tend to be subdivided into separate ones adapted to different plant zones, instead of graded series showing no sudden differences. It also means that animal communities are made much more distinct from one another than would be the case if they were all living in one continuous gradient in conditions, or in a series of open associations of plants like arctic fjaeldmark (stony desert). It would be infinitely more difficult to study animal associations if this were not the case, for we should not have those convenient divisions of the whole fauna into communities which are so useful for working purposes. It is sometimes assumed in discussions on the origin of species that the environmental conditions affecting animals are always in the form of gradients. It is clear that such is by no means always the case.

8. As has been mentioned above, the abrupt transitions between plant communities are particularly well seen on the sides of mountains, where there are vertical zones of vegetation corresponding on a small scale to the big zones of latitude.

In America they are usually referred to as " life zones," and the existence of great mountain ranges in that country is one of the reasons why ecology has attracted more attention there. In England, where the mountain ranges are in the north, we do not see the impressive spectacle of great series of vegetation zones which have so much attracted the American ecologist. This vertical zonation is most striking in the tropics, where, within the same day, one may be eating wild bananas at sea-level and wild strawberries on the mountains. One of the best descriptions of this phenomenon is given by A. R. Wallace in the account of his travels through Java.[57]

The work of botanists has given us fairly clear ideas about the distribution of zones of vegetation, but we are still in great ignorance as to the exact distribution and boundaries of the animal communities in these life zones, and of their relation to the plants. A good deal of work has been done by Americans upon certain groups of animals (chiefly birds and mammals), and in particular may be mentioned the extremely fine account of the Yosemite region of the Sierra Nevada by Grinnell and Storer,[40] in which are given accurate data of the distribution of vertebrates in relation to life zones, together with a mass of interesting notes on the ecology of the animals.

9. There is a further important point in regard to the distribution and composition of animal communities. If we take the community of animals living, say, in the Canadian zone, we should find that a definite percentage are confined to that zone, and in fact that the distribution of some of the animals is strictly determined by the type of vegetation. These species we speak of as " exclusive " to that community. The game-birds found in Great Britain afford good examples of this. The ptarmigan (*Lagopus mutus*) lives in the alpine zone of vegetation, while the red grouse (*Lagopus scoticus*) replaces it at lower levels on the heather moors. Another bird, the capercaillie (*Tetrao urogallus*), lives in coniferous woods, while the pheasant (*Phasianus colchicus*) occurs chiefly in deciduous woods. Finally, the common partridge (*Perdix perdix*) comes in cultivated areas with grassland, etc. We see here examples of birds which are exclusive to certain plant

associations, and we may also note that they are all living a similar life as regards food and general habits. Each association has some kind of large vegetarian bird, although the actual species is different in each case. Another well-known example is the common grass-mouse (*Microtus agrestis*) which, except when it is extremely abundant and " boils over " into neighbouring habitats, is chiefly found living underground in grassland, where it feeds on the roots of the grass. A great number of vegetarian insects are attached to one species of plant, and if that plant only occurs in one association, the animal is also limited in the same way. The oak (*Quercus robur*) supports hundreds of insects peculiar to itself, and if we include parasites the number will be far greater.

10. Continuing our survey of one zone, we should find that there are certain species which occur in particularly large numbers there, although they are not exclusively confined to it. These we call " characteristic " species. A good example of this type is the long-tailed mouse (*Apodemus sylvaticus*) which occurs in woods, but is not confined to them. Trapping data for one area near Oxford showed that 82 per cent. of specimens were caught in woods, while 18 per cent. occurred outside in young plantations and even occasionally in the open. Here the animal is not so strictly limited to one habitat as, for instance, *Microtus*, but we are quite justified in calling it a wood-mouse in this district. Thorpe [122] has described some of the exclusive and characteristic British birds with reference to plant associations.

In most cases in which we have any complete knowledge (and they are few) it is found that these two classes of animals—the exclusives and the characteristics—may often form only a comparatively small section of the whole community, that there are many species of animals which range freely over several zones of vegetation, either because they are not limited by the direct or indirect effects of the vegetation, or because they can withstand a greater range of environment than the others. As an example of this type of distribution we may take the common bank vole (*Evotomys glareolus*) which, in contrast to the *Apodemus* mentioned above, comes both in woodland and in wood margin, shrub communities, and young plantations. The actual figures for comparison with *Apodemus* are as follows : 47 per cent. of the specimens in woods and plantations. 53 per cent. outside, chiefly in shrub or young tree habitats.

It may be, in fact, often rather an arbitrary proceeding to split up the animals living on, say, a mountain-side into communities corresponding to the exclusive and characteristic species of each life zone, and it should be realised quite clearly and constantly borne in mind when doing field work, that many common and important species come in more than one zone. Richards, after several years' study of the animals of an English heath, says : " The commonest animals in a plant community are often those most common elsewhere." [18a] At the same time it is probably true that animals living in several zones of vegetation show a marked tendency to have their limits of distribution coinciding with the edges of the plant zones. This is only natural in view of the step-like nature of the gradient in environment produced by the plant communities.

11. Another important vertical gradient is that found in the sea and in fresh-water lakes, and this is caused by the reduction in the amount of light penetrating the water as the depth increases. This gradient shows itself both in the free-living communities (*plankton*) and in those living on the bottom (*benthos*). As we go deeper down the plants become scarcer owing to lack of light, until at great depths there are no plants at all, and the animals living in such places have to depend for their living upon the dead bodies of organisms falling from the well-lighted zone above, or upon each other. There is the same tendency for the plants to form zones as on land, and one of the most interesting things about marine communities is the fact that certain animals which have become adapted to a sedentary existence compete with the plants (seaweeds of various kinds) and in some cases completely dominate them. The reason for this is that in the sea, and to a lesser extent in fresh water, it is possible for an animal to sit still and have its food brought to it in the water, while on land it has to go and get it. Web-spinning spiders are almost the only group of land animals which has perfected a

means of staying in the same place and obtaining the animals carried along in the air. In the tropics certain big spiders are actually able to snare small birds in their webs. In the sea an enormous number of animals sit still in one place and practically have their food wafted into their mouths. Indeed, food is probably not usually a limiting factor for such animals, and competition is for space to sit on. Hence it is that we find these animals behaving superficially like plants. Over great areas of the tropical seas (dependent probably on certain temperature conditions of the sea, or upon the plankton living in such waters) corals almost completely replace seaweeds on the seashore and shallow waters, where they feed like other animals on plankton organisms or upon small organic particles in the water. Corals on a reef usually form zones, each dominated by one or more species, just as in plants. The zonation is apparently determined by gradients in such factors as surf-action, amount of silt in the water, etc. Amongst the corals grow various calcareous algæ which resemble them very closely in outward appearance. As we go farther from the equator, plants become relatively more and more abundant on the shores of the oceans, but even in Arctic regions certain groups of animals, *e.g.* hydroids, may form zones between other zones composed of seaweeds.[25a]

12. There is another vertical gradient in conditions which is clear-cut and of universal occurrence. This is the gradient in salt content of the water from mountain regions down to sea-level. Through the action of rain all manner of substances are continually being washed out of the rocks and soil. These pass into streams and rivers and accumulate temporarily in lakes and ponds at various levels. But since the salts are always being washed down, we find that on the whole the higher we go the purer is the water. Exceptions must be made to this rule in the case of places which have the higher parts of their mountain ranges composed of very soluble rocks, or in places like Central Asia, which have high plateaux on which many salt lakes develop. But, on the whole, we can distinguish an upland or alpine zone of waters containing few salts and often slightly acid in reaction. Lower down, the

rivers and standing waters contain considerably more salts (*e.g.* places like the Norfolk Broads or the meres of Lancashire and Cheshire). Then there is a rather sudden increase in the steepness of the gradient through brackish water lagoons and estuaries to the sea itself. The sea, having been there

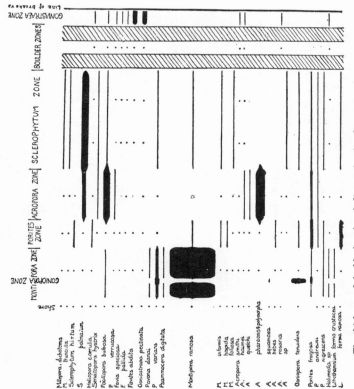

The thickness of the black stripes indicates the abundance of each species at various distances from the shore.

FIG. 1.—Zonation of corals (together with a few calcareous algæ) on a reef in the New Hebrides, showing that the phenomenon of dominance exists among corals, just as among plants. The left-hand side of the diagram is the shore of the island, while the right-hand side is the outer edge of the reef. (From Baker.[104])

much longer than the inland lakes and ponds, contains enormously more salts than the latter, but really it is only one end of a gradient which started high up in the mountains. There are well-marked different associations of animals in all these types of waters. Of course, other factors than salt content are important (particularly temperature), but the salt content itself is undoubtedly very important as a controlling factor, since it acts not only directly but also by affecting the hydrogen ion concentration of the water.

13. Within each of the big zones which owe their existence to major differences in climate there are numerous smaller gradients in outer conditions, each of which gives rise to a series of more or less well-marked associations of plants and animals. These gradients are caused by local variations in soil and climate, or by biotic factors such as grazing by animals. One obvious example is the gradient in the amount of water in the soil. At one end we may find the animal community of a dry heather moor, and at the other the community of free-floating and free-swimming animals which form the plankton of a lake. Between these two extremes there would be zones of marsh, reed swamp, and so on, each with a distinctive set of animals. These various zones are due to the fact that at one end of the gradient there is much soil and practically no water (at any rate in summer), while at the other there is much water and very little soil, the proportion of soil and water gradually changing in between.

14. We can carry the subdivision of animal communities further and split up one ordinary plant association, like an oak wood, into several animal habitats, e.g. tree-tops, tree-trunks, lower vegetation, ground surface, and underground, and we should find that each of these habitats contained an animal community which could be treated to some extent at least as a self-contained unit. Again, each species of plant has a number of animals dependent upon it, and one way of studying the ecology of the animals would be to take each plant separately and work out its fauna. Finally, each animal may contain within its own body a small fauna of parasites, and these again can be split up into associations according to the part of the body which they inhabit. If we examine the parasites of a mouse, for instance, we find that the upper part of the in-testine, the lower part, the caecum, the skin, the ears, each have their peculiar fauna.

It is obviously impossible to enumerate all the different

gradients in the environment and all the different communities of animals which inhabit them. One habitat alone, the edge of a pond, or the ears of mammals, would require a whole book if it were to be treated in an adequate way. The aim of the foregoing sketch of the whole subject is to show that the term " animal community " is really a very elastic one, since we can use it to describe on the one hand the fauna of equatorial forest, and on the other hand the fauna of a mouse's cæcum.

For general descriptions of the animal communities of the more important habitats, the reader may be referred to a book on animal geography by Hesse,[111] and to a more recent book by Haviland.[109]

15. The attention of ecologists has been directed hitherto mainly towards describing the differences between animal communities rather than to the fundamental similarity between them all. The study of these differences forms a kind of animal ethnology, while the study of the resemblances may be compared to human sociology (soon to become social science). As a matter of fact, although a very large body of facts of the first type has been accumulated, few important generalisations have as yet been made from it. So much is this the case that many biologists view with despair the prospect of trying to learn anything about ecology, since the subject appears to them at first sight as a mass of uncoordinated and indigestible facts. It is quite certain that some powerful digestive juice is required which will aid in the assimilation of this mass of interesting but unrelated facts. We have to face the fact that while ecological work is fascinating to do, it is unbearably dull to read about, and this must be because there are so many separate interesting facts and tiny problems in the lives of animals, but few ideas to link the facts together. It seems certain that the key to the situation lies in the study of animal communities from the sociological point of view. This branch of ecology is treated in Chapter V., but first of all it is necessary to say something about the subject of ecological succession— an important phenomenon discovered by botanists, since it enables us to get a fuller understanding of the distribution and relations of animal and plant communities.

CHAPTER III

ECOLOGICAL SUCCESSION

A number of changes (1) are always taking place in animal communities, (2) one of the most important of which is ecological succession, which (3) causes plant associations to move about slowly on the earth's surface, and (4) is partly due to an unstable environment and partly to plant development which typically consists (5) of a *sere* of associations starting with a bare area and ending with a climax association. (6) Each region has a typical set of seres on different types of country which (7) may be studied in various ways, of which the best is direct observation of the changes as in (8) the heather moor described by Ritchie or (9) the changes following the flooding and redraining of the Yser region described by Massart or (10) a hay infusion; but (11) indirect evidence may be obtained as in the case of Shelford's tiger beetles. (12) The stages in succession are not sharply separated and (13) raise a number of interesting problems about competition between species of animals, which (14) may be best studied in very simple communities. (15) In the sea, succession in dominant sessile animals may closely resemble that of land plants, while (16) on land, animals often control the direction of succession in the plants. Therefore (17) plant ecologists cannot afford to ignore animals, while a knowledge of plant succession is essential for animal ecologists.

1. WE have spoken of animal communities so frequently in the last chapter that the reader may be in danger of becoming hypnotised by the mere word "community" into thinking that the assemblage of animals in each habitat forms a completely separate unit, isolated from its surroundings and quite permanent and indestructible. Nothing could be farther from the true state of affairs. The personnel of every community of animals is constantly changing with the ebb and flow of the seasons, with changing weather, and a number of other periodic rhythms in the outer environment. As a result of this it is never possible to find all the members of an animal community active or even on the spot at all at any one moment. To this subject we shall return in the chapter on the Time Factor in Animal Communities, since its

discussion comes more suitably under the structure of animal communities than under their distribution.

2. There is another type of change going on in nearly all communities, the gradual change known as *ecological succession*, and with this we will now deal. Such changes are sometimes huge and last a long time, like the advance and retreat of ice ages with their accompanying pendulum swing from a temperate climate with beech and oak forests and chaffinches, to an arctic one with tundra and snow buntings, or even the complete blotting out of all life by a thick sheet of ice. They may, on the other hand, be on a small scale. Mr. J. D. Brown watched for some years the inhabitants of a hollow in a beech tree, and the ecological succession of the fauna. At first an owl used it for nesting purposes, but as the tissues of the tree grew round the entrance to the hollow it became too small for owls to get into, and the place was then occupied by nesting starlings. Later the hole grew smaller still until after some years no bird could get in, and instead a colony of wasps inhabited it. The last episode in this story was the complete closing up of the entrance-hole. This example may sound trivial, but it is an instance of the kind of changes which are going on continuously in the environment of animals.

3. If it were possible for an ecologist to go up in a balloon and stay there for several hundred years quietly observing the countryside below him, he would no doubt notice a number of curious things before he died, but above all he would notice that the zones of vegetation appeared to be moving about slowly and deliberately in different directions. The plants round the edges of ponds would be seen marching inwards towards the centre until no trace was left of what had once been pieces of standing water in a field. Woods might be seen advancing over grassland or heaths, always preceded by a vanguard of shrubs and smaller trees, or in other places they might be retreating; and he might see even from that height a faint brown scar marking the warren inhabited by the rabbits which were bringing this about. Again and again fires would devastate parts of the country, low-lying areas would be flooded, or pieces of water dried up, and in every

case it would take a good many years for the vegetation to reach its former state. Although bare areas would constantly be formed through various agencies, only a short time would elapse before they were clothed with plants once more.

There are very few really permanent bare areas to be met with in nature. Rocks which appear bare at a distance are nearly always covered with lichens, and usually support a definite though meagre fauna, ranging from rotifers to eagles. Apparently barren places like lakes contain a huge microscopic flora and fauna, and even temporary pools of rain-water are colonised with almost miraculous rapidity by protozoa and other small animals.

4. It is the exception rather than the rule for any habitat to remain the same for a long period of years. Slow geological processes, like erosion and deposition by rivers and by the sea, are at work everywhere. Then there are sudden disasters, like fires, floods, droughts, avalanches, the introduction of civilised Europeans and of rabbits, any of which may destroy much of the existing vegetation. There is a third kind of change which is extremely important but not so obvious, and is the more interesting since its movements are orderly and often predictable. This is the process known as the *development* of plant communities. Development is a term used by plant ecologists in a special technical sense, to include changes in plant communities which are solely or largely brought about by the activities of the plants themselves. Plants, like many animals, are constantly moulting, and the dead leaves produced accumulate in the soil below them and help to form humus. This humus changes the character of the soil in such a way that it may actually become no longer suitable for the plants that live there, with the result that other species come in and replace them. Sometimes the seedlings of the dominant plant (*e.g.* a forest tree) are unable to grow up properly in the shade of their own parents, while those of other trees can. This again leads to the gradual replacement of one community by another.

When a bare area is formed by any of the agencies we have mentioned, *e.g.* the changing course of a river, it is first colonised by mosses or algæ or lichens ; these are driven out by low herbs, which kill the pioneer mosses by their shade ; these again may be followed by a shrub stage ; and finally a woodland community is formed, with some of the earlier pioneers still living in the shade of the trees.

This woodland may form a comparatively stable phase, and is then called a climax association, or it may give way to one or more further forest stages dominated by different species of trees in the manner described above. It is not really possible to separate development of communities from succession caused by extrinsic changes, such as the gradual leaching out of salts from the soil or other such factors unconnected with the plants themselves. The important idea to grasp is that plants react on their surroundings and in many cases *drive themselves out*. In the early stages of colonisation of bare areas the succession is to a large extent a matter of the time taken for the different plants to get there and grow up ; for obviously mosses can colonise more quickly than trees.

5. In any one region the kind of climax reached depends primarily upon the climate. In high Arctic regions succession may never get beyond a closed association of lichens, containing no animals whatsoever. In milder Arctic regions a low shrub climax is attained, while farther south the natural climax is forest or in some cases heath, according to whether the climate is of a continental or an oceanic type. Sometimes ecological succession is held up by other agencies than climate and prevented from reaching its natural climax. In such cases it is a common custom to refer to the stage at which it stops as a sub-climax. A great deal of grassland and heath comes under this heading, for further development is prevented by grazing animals, which destroy the seedlings of the stage next in succession. An area of a typical heather moor in the New Forest was fenced off for several years from grazing ponies and cattle by its owner, with the immediate result that birches and pines appeared by natural colonisation, and the young pines, although slower in growth than the birches, will ultimately replace them and form a pine wood. Here grazing was the sole factor preventing normal ecological

succession. The same thing is well known to occur in a great many places when heather or grass is protected, the important animals varying in different places, being usually cattle, sheep, horses, or rabbits, or even mice.

6. We begin to see that the succession of plant communities does not take place at random, but in a series of orderly stages, which can be predicted with some accuracy. The exact type of communities and the order in which they replace one another depend upon the climate and soil and other local factors, such as grazing. It is possible to classify different series of stages in succession in any one area, the term "sere" being used to denote a complete change from a bare area in water or soil up

to a climax like pine wood (*cf.* Fig 2). Each type of soil, etc., has a different type of "sere" which tends to develop upon it, but they all have one character in common: bare areas are usually very wet or very dry, and the tendency of succession is always to establish a climax which is living in soil of an intermediate wetness—a type of vegetation called "mesophytic," of which a typical example is an oak wood. Thus a dry rock surface gains ultimately a fairly damp soil by the deposition of humus, while a water-logged soil is gradually raised above the water-level by the same agency, so that there tends to appear a habitat in which the expenditure by plants and by direct loss from the soil is suitably balanced by the

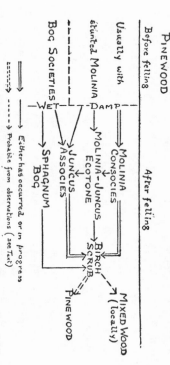

FIG. 2.—The diagram shows the stages in ecological succession following colonisation of damp bare areas formed by felling of a pine wood on Oxshott Common. The succession is different on the drier areas. (From Summerhayes and Williams.[120])

income of water, and extremes of environment are avoided. This is, of course, only a rough generalisation and applies especially to temperate regions, but it explains why we often get seres on very different kinds of bare areas converging towards the same final climax.

7. The account of this subject given above is necessarily brief, and a much fuller account is given by Tansley in his book *Practical Plant Ecology*,[15] which is essential to the work of all animal ecologists. Clements has treated the whole subject in stupendous detail in his monograph *Plant Succession*,[39] which is illustrated by a very fine series of photographs of plant communities.

Let us now consider a few examples of succession in animal and plant communities. It is clearly impracticable to take more than a few species as examples of changes in whole communities, and naturally the exclusives afford the most striking ones. There are several ways in which animal succession can be studied. The best way is to watch one spot changing over a series of years and record what happens to the fauna. This is the method least practised, but the most likely to lead to productive results, since we stand a good chance of seeing how the structure of the communities is altered as one grades into another. Yapp [31] says : "We may perhaps regard the organisms, both plants and animals, occupying any given habitat, as woven into a complex but unstable web of life. The character of the web may change as new organisms appear on the scene and old ones disappear during the phases of succession, but the web itself remains." It is just the changes in this "web" about which we know so little at present, and that is why study of the actual changes will always be the most valuable.

8. One of the most interesting and clear-cut examples of succession, recorded by Ritchie,[13a] is so striking that it has been often quoted, and is worth quoting again here. He describes the manner in which a typical heather moor in the south of Scotland, with its normal inhabitant, the red grouse (*Lagopus scoticus*), was converted in the short space of fifteen years into a waste of rushes and docks, inhabited by a huge

colony of black-headed gulls (*Larus ridibundus*), and then in about ten years turned back again into a moor like the original one. These events were brought about by the arrival of a few pairs of gulls which nested there for the first time in about 1892. The gulls were protected by the owner, and after fifteen years they had increased prodigiously until there were well over 3,000 birds nesting. The occupation of the ground by gulls, with its accompanying manuring and trampling of the soil, caused the heather to disappear gradually and to give way to coarse grass. The grass was then largely replaced by rushes (*Juncus*), and the latter ultimately by a mass of docks (*Rumex*). At the same time pools of water formed among the vegetation and attracted numbers of teal (*Anas crecca*). The grouse meanwhile had vanished. Then protection of the gulls ceased, and their numbers began to decrease again, until in 1917 there were less than sixty gulls nesting, the teal had practically disappeared, and the grouse were beginning to return. In fact, with the cessation of "gull action" on the ground the place gradually returned to its original state as a heather moor. As Ritchie remarks, there must have been a huge number of similar changes among the lower animals which also would be profoundly affected by the changes in vegetation.

9. Another striking story is that told us by Massart,[35] who studied the changes wrought during the war by the flooding of parts of Belgium in the Yser district. Here the sea was allowed by the Belgian engineers to inundate the country in order to prevent the advance of the German army. The sea-water killed off practically every single plant in this district, and all available places were very soon colonised by marine animals and plants, space being valuable in the sea. When the country was drained again at the end of the war, ecological succession was seen taking place on a generous scale. At first the bare "sea-bottom" was colonised by a flora of salt-marsh plants, but these gave way gradually to an almost normal vegetation until in many places the only traces of the advance and retreat of the sea were the skeletons of barnacles (*Balanus*) and mussels (*Mytilus*) on fences and notice-boards, and the presence

of prawns (*Palemonetes varians*) left behind in some of the shell holes.

10. Ecological succession may easily be studied experimentally by making a hay infusion in water and leaving it exposed to the air for several weeks. Bacteria are the first things to become abundant, since they live upon the decaying vegetable matter. Then various protozoa appear, and it is possible to see a whole animal community being gradually built up, as each new species arrives and multiplies and fits into its proper niche. In a hay infusion the bacteria are followed by small ciliate protozoa of the *Paramecium* type, which subsist upon bacteria and also by absorbing substances in solution and in suspension in the water. Then there are larger hypotrichous ciliates, which prey upon bacteria and also upon the smaller ciliates. Eventually the whole culture may degenerate owing to the exhaustion of food material for the bacteria and therefore for the animals dependent on them. On the other hand, green plants, in the form of small algæ, may arrive and colonise the culture. These will be able to subsist for a long time, and may change the character of the whole community by providing a different type of food. Succession in hay infusions is particularly fascinating, since it can be studied anywhere, and does not last over a very long time.

11. Another method of determining the course of animal succession is to work from a knowledge of the succession relations of the plant communities (gained either from direct observation or from deduction and comparison with other districts) and then work out the animal communities of each plant zone. It is then possible to say in a general way what animals will replace existing ones when succession does occur. This was done by Shelford,[20] who studied the tiger beetles of the genus *Cicindela* (carnivorous ground beetles of variegated colours) in a sere of plant communities on the shores of Lake Michigan. The lake-level has been falling gradually of late years, and there can be seen all stages in succession on the bare areas left behind on the shores. On the lake margin was *Cicindela cuprascens*, whose larvæ live in wettish sand. Young

cottonwoods colonise this ground, and another species of tiger beetle (*C. lepida*) then replaces the first one. In the old cotton-woods, where conditions are different with grass and young pine seedlings. *C. formosa* took the place of *C. lepida*. Then with the formation of a dominant pine community on the ridges still another species, *C. scutellaris*, replaces the previous one and continues to live on into the next stage in succession, a black oak community; but it gradually becomes scarcer with the change to white oak, until a stage is reached with no tiger beetles at all. With the following red oak stage there arrives *C. sexguttata*, which persists afterwards in clearings, but when the climax association of beech and maple has been reached tiger beetles again disappear altogether. Here there can be distinguished at least eight stages in the development of plant associations, and five different species of tiger beetle, none of which come in more than two plant zones.

12. It is important to note that we have to deal in this case with a genus of animals which tends to form species which are exclusive or confined to one or two plant associations. In England there seems to be the same tendency among the species of *Cicindela*. But this kind of strict limitation to plant associations is probably rather unusual, especially among carnivorous animals. Exception must be made in the case of some of the great host of herbivorous animals (in particular insects) which are attached to one species of plant only. But even in these cases the plant itself, and therefore the animal, is not usually confined to one plant association. In practice, succession in animal communities is an infinitely more com-plicated affair. One reason for this is the great lag of animals behind plants, due to their different powers of dispersal. Another reason is that the survival of only a few of the earlier plants in the later stages will enable a great number of animals to hang on also, and these of course cannot be separated in their inter-relations with the newcomers. For instance, on areas in the south of England where pine woods have grown up over and largely replaced heather (*Calluna*), there are still many patches of heather growing in the more open places, and these have been found to contain the typical heather

communities of animals, showing that the animals are in this case affected rather by the food, shelter, etc., provided by the heather than by the general physical " climate " produced by the pine wood.[18c] It is when we try to work out the food relations of the animals that the presence of small patches of earlier pioneer animals in a climax association becomes such a complicating factor. In fact, succession (at any rate in animals) does not take place with the beautiful simplicity which we could desire, and it is better to realise this fact once and for all rather than to try and reduce the whole phenomenon to a set of rules which are always broken in practice ! The present state of our knowledge of succession is very meagre, and this ignorance is to a large extent due to lack of exact knowledge about the factors which limit animals in their distribution and numbers. The work done so far has necessarily been restricted to showing the changes in exclusive species of one genus, or in the picking out of one or two salient features in the changes as an indication of the sort of thing that is taking place. Work like that done by Shelford, and observations like those of Ritchie and Massart, make it quite clear that succession is an important phenomenon in animal life ; the next stage of the inquiry is the discovery of the exact manner in which succession affects whole communities.

13. It will be as well at this point to remind the reader that most of the work done so far upon animal succession has been static and not dynamic in character ; that the cases in which the whole thing has been seen to happen are few in number and, although extremely valuable and interesting, of necessity incompletely worked out.

Given a good ecological survey of animal communities and a knowledge of the local plant seres, we can predict in a general way the course of succession among the animals, but in doing so we are in danger of making a good many assumptions, and we do not get any clear conception of the exact way in which one species replaces another. Does it drive the other one out by competition ? and if so, what precisely do we mean by competition ? Or do changing conditions destroy or drive out the first arrival, making thereby an empty niche for another animal which quietly replaces it without ever becoming " red in tooth and claw " at all ? Succession brings the ecologist face to face with the whole problem of competition among animals, a problem which does not puzzle most people because they seldom if ever think out its implications at all carefully. At the present time it is well known that the American grey squirrel is replacing the native red squirrel in various parts of England, but it is entirely unknown why this is occurring, and no good explanation seems to exist. And yet, more is known about squirrels than about most other animals. In ecological succession among animals there are thousands of similar cases cropping up, practically all of which are as little accounted for as that of the squirrels. There is plenty of work to do in ecology.

14. It is probable that accurate data about the succession in animal communities will be most easily and successfully obtained by taking very small and limited communities living in peculiar habitats, for it is here that the number of species is reduced to reasonable proportions. Experience has shown that the general study of animal communities is best carried out on simple communities such as those of Arctic regions or of brackish water. Habitats in which succession can be studied quickly and conveniently are the dead bodies of animals, the dead bodies of plants (*e.g.* logs, or fungi), the dung of mammals, marine timber, temporary pools, and so on. The ecologist will be able to find a number of such habitats wherever he is ; and they all contain on a small scale similar communities to those found in woods or lakes. In such places succession is always in progress and, what is more important, in quick progress. The writer has found that it is almost impossible to make even a superficial study of succession in any large and complicated community, owing to the appalling amount of mere collecting which is required, and the trouble of getting the collected material identified. When one has to include the seasonal changes throughout the year as well, the work becomes first of all disheartening, then terrific, and finally impossible. Much of this mental strain can be avoided by choosing simple communities, and the results as contributions to the general

theory of succession are probably just as valuable. In any case, it is desirable that botanists and animal ecologists should cooperate in such studies, and in most cases a team of several people is required for the proper working out of the animals. It is probable, however, that in the simpler cases one man could do very valuable work.

15. We have not spoken so far of succession in the sea. Owing to the peculiar importance of sessile animals on the shores and slopes of the sea, it is not uncommon to find plant and animal succession becoming rather closely related. Wilson [82] watched the succession on bare areas on the shores of California at La Jolla, and found that the pioneers were colonial diatoms which formed the first community. These were followed by an association of colonial hydroids (mostly *Obelia*), and the latter were then replaced by a seaweed (*Ectocarpus*), which became dominant after some four months. Further stages were foreshadowed, and it appeared that the whole would ultimately develop into a climax association of other seaweeds (chiefly kelp). The interesting thing about this sere is the fact that the second stage in succession is formed by sessile animals, and is sandwiched between two plant stages. The zonation of animals and plants in the inter-tidal zone in other regions is often an alternation of animal and plant dominance, *e.g. Balanus* or *Mytilus* and seaweeds, in the temperate regions. On coral islands succession may consist almost entirely of a series of animal zones with only an occasional plant zone formed of calcareous algæ.[104] In such cases it is perfectly legitimate to refer to the most abundant animal as the " dominant " species, but in most animal com-munities the term has little meaning owing to the different methods of feeding adopted by animals and plants. It is perhaps better to avoid the use of the term " dominant " in the cases of animals except for sessile aquatic species, since dominance implies occupying more space or getting more light than other species.

16. We have just pointed out that animals play an important part in ecological succession in the sea ; but it should also be realised that they also have very important effects in a different

sort of way upon plant succession on land. Herbivorous animals are often the prime controlling factor in the particular kind of succession which takes place on an area. Farrow [8a] says : " It is thus seen that variation in the intensity of the rabbit attack alone is sufficient to change the dominant type of vegetation in Breckland from pine woodland to dwarf grass-heath vegetation through the phases of *Calluna* heath and *Carex arenaria*, and that for each given intensity of rabbit attack there is a certain associated vegetation." He showed that there were definite zones of vegetation round each rabbit colony and even each hole, since the distance from the colony resulted in different intensities of attack. The curious fact also appears that *Carex arenaria* will dominate *Calluna* when both are eaten down intensely. It is the relative intensity of attack that matters. The example quoted here is only one of many which could be given, although the situation has not been worked out so ingeniously or fully for any other animals or place.

17. Hofman's work [83] has made it probable that the influence of rodents in burying seeds of conifers may sometimes determine the type of succession which starts after a forest fire. The Douglas Fir (*Pseudotsuga taxifolia*) is a dominant tree over large parts of the Cascade and Coast region of Washington and Oregon. When the seed crops are light they are largely destroyed by an insect (*Megastigmus spermotrophus*) and by rodents. But when there is an unusually heavy crop large quantities of surplus seed are collected and stored in caches by the rodents. Since in many cases the animals do not return to the caches the seeds remain there for a good many years, and if there is a fire in the forest or the trees are cut down, large quantities of the stored seed germinate. More of the Douglas fir seed than of other species of trees such as hemlock and cedar is cached by rodents, so that the Douglas has accordingly an advantage in the early stages of succession. The same thing holds good for the white pine ; its seeds are much eaten by rodents, which gather them and store them in the ground. But here other factors such as germination come in and affect the succession.

18. One more example may be given of the way in which animals and plants are intimately bound up together in ecological succession. Cooper [84] has studied the relation of the white pine blister rust to succession in New England and the Adirondacks. This disease is a very important one, and the most critical point in its life-history, from the point of view of controlling it, is its occurrence at one stage on the various species of wild gooseberries (*Ribes*). Cooper found that the distribution of *Ribes* was to an important extent affected by the distribution of fruit-eating birds, and that changes in the character of the birds during ecological succession resulted in a failure of the gooseberry seeds to spread in sufficient numbers to establish many new plants, after a certain stage in succession had been reached. The normal succession in places where the original forest had been cleared and then allowed to grow up again is as follows : first a stage with rank grass and weeds ; this is followed by a shrub stage, in which plants like raspberry, blackberry, juniper, etc., are important. At this point the species of *Ribes*, though relatively unimportant, reach their maximum abundance. The shrub stage is followed by one with trees, of which the most important are white pine, aspen, birch, and maple. Finally a climax of other species may be reached. But the important point is the appearance of the first trees, for at this stage the bird fauna changes considerably and the number of fruit-eating, or at any rate gooseberry-eating, birds diminishes suddenly. The *Ribes* is able to exist in the shade of these later forest stages of succession, but cannot produce fruit in any quantity, and so, unless birds bring in new seeds from outside, the *Ribes* is bound to die out ultimately. As we have seen, the corresponding changes in the bird species prevent any large amount of seed getting into the forest after these stages, and this reacts upon the rust.

19. It is obvious that a knowledge of animals may be of enormous value to botanists working on plant succession. At the same time it is necessary for the animal ecologist to have a good general knowledge of plant succession in the region where he is working. It enables him to classify and

understand plant communities, and therefore animal communities, much more easily than he can otherwise do, by providing an idea which links up a number of otherwise rather isolated facts. Furthermore, as will be shown later on, succession of communities, which is nothing more than a migration of the animals' environment, plays an important part in the slow dispersal of animals. Again, as we have pointed out, the intricate problems of competition between different species of animals can be studied to advantage in a series of changing communities. Tansley's *Types of British Vegetation* [16] is the standard work in which the plant associations of Britain are described, and in many cases the probable lines of succession in the associations are given also. It is essential to have access to this book if one is carrying out any extensive work in the way of preliminary surveys or studies in ecological succession.

CHAPTER IV

ENVIRONMENTAL FACTORS

The ecologist is (1) concerned with *what animals do*, and with the factors which prevent them from doing various things : (2) the study of factors which limit species to particular habitats lies on the borderlines of so many subjects, that (3) he requires amongst other things a slight knowledge of a great many scattered subjects. (4), (5), (6), (7) The ecology of the copepod *Eurytemora* can be taken as a good example of the methods by which such problems may be studied, and illustrates several points, *e.g.* (8) the fact that animals usually have appropriate psychological reactions by which they find a suitable habitat, so that (9) the ecologist does not need to concern himself very much with the physiological limits which animals can endure ; and (10) the fact that animals are not completely hemmed in by their environment, but by only a few limiting factors, which (11) may, however, be difficult to discover, since (12) the factor which appears to be the cause may turn out to be only correlated with the true cause. (13), (14) Examples of limiting factors to the distribution of species are : hydrogen ion concentration of water ; (15) water supply and shelter ; (16) temperature ; (17) food plants ; and (18) interrelations with other animals. The last subject is so huge and complicated, that it requires special treatment— as the study of animal communities.

1. IT will probably have occurred to the reader, if he has got as far as this, that rather little is known about animal ecology. This is, of course, all to the good in one way, since one of the most attractive things about the subject is the fact that it is possible for almost any one doing ecological work on the right lines to strike upon some new and exciting fact or idea. At the same time it is often rather difficult to know what are the best methods to adopt in tackling the various problems which arise during the course of the work. This is particularly true of the branch of ecology dealt with in the present chapter. Much of the work that is done under the name of ecology is not ecology at all, but either pure physiology—*i.e.* finding out how animals work internally—or pure geology or meteorology, or some other science concerned primarily with the

outer world. In solving ecological problems we are concerned with *what animals do* in their capacity as whole, living animals, not as dead animals or as a series of parts of animals. We have next to study the circumstances under which they do these things, and, most important of all, the limiting factors which prevent them from doing certain other things. By solving these questions it is possible to discover the reasons for the distribution and numbers of different animals in nature.

2. It is usual to speak of an animal as living in a certain physical and chemical environment, but it should always be remembered that strictly speaking we cannot say exactly where the animal ends and the environment begins—unless it is dead, in which case it has ceased to be a proper animal at all; although the dead body forms an important historical record of some of the animal's actions, etc., while it was alive. The study of dead animals or their macerated skeletons, which has tended to form such an important and necessary part of zoological work, and which has bulked so largely in the interest of zoologists for the last hundred years, has tended to obscure the important fact that animals *are* a part of their environment. There are numerous gases, liquids, and solids circulating everywhere in nature, the study of which is carried on by physicists, chemists, meteorologists, astronomers, etc.; certain parts of these great systems are, as it were, cut off and formed into little temporary systems which are animals and plants, and which form the objects of study of physiologists and psychologists. Ecological work is to a large extent concerned with the interrelations of all these different systems, and it must be quite clear that the study of the manner in which environmental factors affect animals lies on the borderline of a great many different subjects, and that the task of the ecologist is to be a sort of liaison officer between these subjects. He requires a slight, but not superficial, knowledge of a great many branches of science, and in consequence must be prepared to be rather unpopular with experts in those sciences, most of whom will view him with all the distaste of an expert for an amateur. At the same time, although his knowledge of these other subjects can only be slight, owing to the absolute impossibility of

learning them thoroughly in the time at his disposal, he has the satisfaction of being able to solve problems, often of great practical importance, which cannot possibly be solved by having a profound grasp of only one field of science. Many people find it rather a strain if their work includes more than one or two " subjects," and this is probably the reason why many who start doing ecological work end up by specialising on one of the ready-made subjects with which they come in contact in the course of their ecological researches. It is quite true that one is frequently held up by the absence of existing data about some problem in geology or chemistry, and it may be necessary to turn to and try to solve it oneself. It is also true that the division of science into water-tight compartments is to be avoided like the plague. At the same time there are problems in the reactions of animals with their environment which call for a special point of view and a special equipment, and one of the most important of these is a slight knowledge of a number of different subjects, if only a knowledge of whom to ask or where to look up the information that is required.

3. Suppose one is studying the factors limiting the distribution of animals living in an estuary. One would need to know amongst other things what the tides were (but not the theories as to how and why they occur in a particular way) ; the chemical composition of the water and how to estimate the chloride content (but not the reasons why silver nitrate precipitates sodium chloride) ; how the rainfall at different times of the year affected the muddiness of the water ; something about the physiology of sulphur bacteria which prevent animals from living in certain parts of the estuary ; the names of common plants growing in salt-marshes ; something about the periodicity of droughts (but not the reasons for their occurrence). One would also have to learn how to talk politely to a fisherman or to the man who catches prawns, how to stalk a bird with field-glasses, and possibly how to drive a car or sail a boat. Knowing all these things, and a great deal more, the main part of one's work would still be the observation and collection of animals with a view to finding out their distribution and habits. Having obtained

the main data about the animals, one would be faced by a number of inexplicable and fascinating cases of limited distribution of species, upon which would have to be focussed every single scrap of other knowledge which might bear on the problems. Having crystallised the problems it is often desirable to carry out a few simple experiments; but more often, careful observation in the field will show that the experiment has already been done for one in nature, or by some one else—unintentionally.

4. Let us take an example which will give some idea of the way in which the factors limiting the distribution of animals may be studied. To continue the same line of thought as before, we will consider *Eurytemora lacinulata*, a copepod crustacean which normally inhabits weak brackish water in lagoons and estuaries. In an estuarine stream near Liverpool which has been studied, *E. lacinulata* is absent from the upper fresh-water parts, as also from fresh-water ponds in the neighbourhood, but begins to occur in the region which is influenced for a short time only by salt water at the height of the tides. In this part of the stream the environment consists of fresh water at low tides, brackish water (e.g. up to a salinity of 4 (chlorides) per mille) at high tide, with a gradation in between. Further downstream, under more saline conditions, the *Eurytemora* again disappears. Near this stream there is a moat surrounding an old duck decoy, and filled with brackish water whose salinity fluctuates slightly about 4 per mille, and in which *Eurytemora* lives and flourishes. It is clear that in this region *E. lacinulata* lives in weak brackish water up to about 4 per mille, or in alternating weak brackish and fresh water, but neither in permanent fresh water nor in more saline water. We are justified in making the hypothesis that *E. lacinulata* requires a certain optimum salinity, any wide deviation from which is unfavourable to it.

5. Now it has been found, on the other hand, that in the Norfolk Broads this species lives in the fresh-water broads and is almost entirely replaced in the brackish ones by an allied species, *E. affinis*. Furthermore, there are a certain number of curious records of the occurrence of *E. lacinulata*

in inland fresh-water ponds which contain no unusual amount of salt in the water. For instance, it lives permanently in a small pond in the Oxford Botanic Gardens. Therefore fresh water by itself is no bar to the survival of this animal (if we assume, as we are doing here, that there are not several different physiological races of this species, which have different requirements in the matter of salinity). What is the cause of its absence from the fresh-water parts of the Liver-pool stream? The chief way in which the fresh and brackish parts of the stream differ, apart from the salinity, is in flora and fauna, the upper part having different associations of animals from the lower. So perhaps there is some biotic limiting factor. This idea is supported by outside evidence. If we examine the cases mentioned above of its occurrence inland, we notice several remarkable things about them. In the first place, the ponds in which *Eurytemora* occurs are few, and these few are all artificial and fairly new, mostly under twenty years of age. In the second place, it is absent from all older ponds, many of which are very similar to the newer ones in fauna, flora, and general conditions. But these older ponds do differ in one important respect : they all contain another copepod, *Diaptomus gracilis*, which can be shown to have bad powers of dispersal, and which therefore does not reach a new pond for some time, usually not less than thirty or forty years. The ponds containing *Diaptomus* have in them no *Eurytemora*, and the ponds containing *Eurytemora* have no *Diaptomus*. We can explain the latter fact by the bad dispersal powers of the *Diaptomus*, but how are we to explain the former ? It appears that there must be some form of competition (see p. 27) between the two species, in which *Diaptomus* is success-ful. At this point we turn for further evidence to an experiment carried out with great care (although unintentionally) by the Oxford Corporation. It is always exhilarating to find an experiment already done for you on a large scale, and in nature, instead of having to do it yourself in a laboratory. There is a large pond in the Oxford waterworks in which *Diaptomus gracilis* occurs abundantly, but from which *Eurytemora* is absent. A few yards away there is a series of filter-beds which are dried out and cleaned every few weeks to remove the conferva which grows upon them. *Diaptomus* does not live in the filter-beds because its eggs cannot withstand being dried up, but there are swarms of *Eurytemora lacinulata* taking its place. In this case the possibility of dispersal causing the different distribution is removed, and we have practically a controlled experiment.

6. The upshot of all this is that *E. lacinulata* appears to be limited in its inland distribution to ponds which do not contain *Diaptomus*. Returning to the original stream near Liverpool, it now seems very likely that there are biotic factors at work limiting the distribution of the *Eurytemora*, since we have knocked out salinity as a possibility, and shown that competition is probably an important factor elsewhere. Now, as *Diaptomus* does not occur in the stream, some other animal must be acting as limiting factor or competitor. So far the inquiry has not been pushed further, but the example has been given in order to show that it pays to make the investigation as wide as possible in order to find what is probably the best line to follow up. In this case one would concentrate upon the animal interrelations of *Eurytemora*, knowing that there was a good chance of solving the problem along that line.

7. There would still remain the question of the limiting factor or factors of *Eurytemora* at the other end of its range. This has not been studied on *E. lacinulata*, but some work has been done upon *E. raboti*, and this will serve to illustrate methods, although of course no conclusions based on one species must ever be applied to another. (There have been not infrequently cases in which some one has repeated a man's work in order to confirm his experiments, using for his material not the original species but a closely allied one, and has been able to contradict the conclusions of the first man—not unnaturally, when we realise how enormously closely allied species may differ in their physiological reactions.)

Eurytemora raboti inhabits weak brackish water in tidal lagoons on the coasts of Spitsbergen, and sometimes when these become cut off from the sea by rising of the land, the copepods survive successfully in the relict lagoons, which then contain only fresh water. The only factor preventing this species from living in other fresh-water ponds appears to be bad powers of dispersal, or at any rate difficulty in establishing itself from a few eggs carried by accidental dispersal. At the other end of its habitat it disappears at a certain point where the salinity at high tides is over about 9 per mille (chlorides). To find out whether high salinity itself was the limiting factor an experiment was tried, and it was found that the animals died in a salinity of 13 per mille after about two hours, while they were quite unharmed in a salinity of 9 per mille after more than three hours. The experiment showed with some certainty that salinity was in fact the cause of its limited distribution in estuarine regions.[25b]

8. When one is studying limiting factors, it is really more important to have a nodding acquaintance with some of the things which are going on in the environment, than to know very much about the physiology of the animals themselves. This statement may sound odd, and is certainly rather in opposition to much of the current ecological teaching, but there is a perfectly good reason for making it. Most animals have some more or less efficient means of finding and remaining in the habitat which is most favourable to them. This may be done by a simple tropism or by some elaborate instinct. Examples of animals which employ the first type of response are the white butterflies *Pieris rapæ* and *brassicæ*, which select the leaves of certain species of cruciferous plants for the purpose of laying their eggs, in response to the chemical stimulus of mustard oils contained in the leaves of the plants.[105] An example of the complicated type is the African lion, which chooses its lair with great attention to a number of different and rather subtle factors. It is usually supposed that animals choose their habitats merely by avoiding all the places which are physiologically dangerous to them, in the same way that a *Paramecium* turns away from certain kinds of chemical stimuli in the water in which it is swimming. This is true in one sense ; but the stimuli which lead an animal to keep away from the wrong habitat are not usually capable of doing any direct harm to it, and are much more in the nature of warning

signals which indicate to it that if it goes much further into this unsuitable habitat, or remains there too long, the results will be dangerous. For instance, the fresh-water amphipod *Gammarus pulex* is a nocturnal animal, which lives under stones and plants during the day, its habitat being determined by a simple negative reaction to light. But light is not in itself harmful to Gammarus (except possibly in a very early stage of development), and the reason for living in the dark is almost certainly because of the protection which it gets from enemies, since the animal has no effective means of defending itself against enemies except by swimming away or hiding. Often, however, the signs by which animals choose their habitats are not warnings of danger to the animal itself, but have the effect of keeping it out of places in which it could not breed or bring up its young successfully.

9. The point which we are trying to make is that most animals are, in practice, limited in their distribution by their habits and reactions, the latter being so adjusted that they choose places to live in, which are suitable to their particular physiological requirements or to their breeding habits. The latter may often be much more important than the former: the willow wren, *Phylloscopus trochilus*, and the chiffchaff, *P. collybita*, both range through similar kinds of vegetation, and do not appear to be affected directly by the physical or biotic conditions at different levels of a wood or in different plant associations. At the same time, the habitats for breeding are very markedly different, the willow wren choosing places where the ground vegetation is low, while the chiffchaff occurs in woods which have grown up and in which the undergrowth has correspondingly changed.[11a] We might put the matter this way: every animal has a certain range of external conditions in which it can live successfully. The ultimate limits of environment are set by its physiological make-up; if these limits are reached the animal will die. It is therefore undesirable that the animal should run the risk of meeting such dangerous conditions, and it has various psychological reactions which enable it to choose, to a large extent, the optimum conditions for life. The animal is not

usually occupying the extreme range of conditions in which it could survive, since at the limits it is not so efficient, and because the actual habitat is usually still further limited by the breeding requirements. In other words, it is usually possible to use the psychological reactions of animals as an indication of their physiological " abilities," and to that extent it is possible to solve ecological problems without knowing a great deal about the physiological reasons *why* certain conditions are unsuitable. We simply assume that they are unsuitable from the fact that animals avoid places where they occur. (This adjustment is presumably brought about by the process of natural selection acting over very long periods, since animals which chose a habitat which turned out to be unsuitable would inevitably die or fail to breed successfully unless they could leave it in time.) By making the assumption that animals are fairly well adapted to their surroundings we certainly run a risk of making serious mistakes in a few cases, because owing to the lag in the operation of natural selection (see p. 185) animals are not by any means always perfectly adapted to their surroundings. But the rule is useful in a general way, and may prevent an ecologist from getting sidetracked upon purely physiological work which, however interesting and valuable in itself, does not throw any light on the ecological problems which he set out to solve.

10. We may conclude this general discussion of the relation of animals to their environment by referring more particularly to the important general idea of limiting factors. Animals are not completely hemmed in by their environment in any simple sense, but are nearly always prevented from occupying neighbouring habitats by one or two limiting factors only. For instance, in the example given above, *Eurytemora lacinulata* was quite capable of living in fresh water, but was prevented from doing so by some other quite different factor. It was adapted to live in low salinities, but was unable to realise this power completely. This may be said to apply to all animals. They are limited in any one direction by one or two factors, but are otherwise quite able to survive a wider range of conditions. Crayfish would be able to live successfully in many rivers

where they are not found at present if the latter contained a sufficiently high content of calcium carbonate, a substance which the crayfish needs for the construction of its exoskeleton.[106]   In this case, a study of the oxygen requirements of the crayfish might throw practically no light on the reasons for its distribution.   There are many instances of herbivorous animals which follow their food-plant into widely different climates from those to which they are accustomed, when the plant itself spreads into a new country.   It may be taken as a rule that animals are never fully utilising all their possibilities, owing to the presence of a few limiting factors. In fact, an animal is limited by the things at which it is least efficient, and if these disabilities are removed it can immediately occupy a new range of habitat.   We might put this in another way and say that in order to occupy a new environment the animal has only to alter one or two of its psychological or physiological characteristics.   This idea is of some importance from the point of view of theories of evolution and adaptation.

11. We may now consider some examples of the way in which environmental factors limit the distribution of wild animals.   Very frequently when a particular species is being studied, it is found that the distribution of the animal is correlated with some definite feature in its environment, but it must not be assumed from this evidence that that factor is the actual limiting one which is at work.   This was clearly seen in the case of *Eurytemora*, in which salinity at first sight appeared to be the limiting condition in its distribution, whereas the factor was really a different one which happened to be correlated with the salinity.   This principle is of very wide application in ecological work.   The habitat of any animal can only be accurately described in terms of actual limiting factors, but in practice we have to describe it in a very rough way by means of other factors which are correlated with the really important ones.   When we say that an amphibian lives in caves, that is only a way of indicating that it lives under a set of environmental conditions of light, heat, and food, which can be roughly described as cave conditions.   In such a case careful investigation might show that the real determining

factor was something quite different from any of these, *e.g.* high lime-content of the water. There are so many environmental factors at work, and these are so often closely interwoven in their action that it may be difficult to say exactly which one is playing the vital part in determining an animal's distribution. For instance, the humidity of the air depends upon a number of things, such as the temperature, wind, and rainfall, while these in turn may be controlled by the type of vegetation or the degree of exposure.

Suppose we take *Silpha quadripunctata*, a beetle which occurs in England almost entirely in oak woods. Since it is carnivorous, the factors which actually limit it to this type of wood might be any out of a very large selection, *e.g.* food, humidity, temperature, the right colour of background, or breeding conditions involving the same or other factors. Again, the northern limit of the mouse *Apodemus sylvaticus* in Norway coincides roughly with latitude 62° N., in other words, certain temperature conditions. But another factor which is correlated also with latitude is the length of the night, and since this species of mouse is nocturnal it might be very well limited by this factor, since in latitude 62° the night only lasts five hours in the height of summer, and there is consequently only a very short period of darkness in which the mouse can feed. Without further work upon the general ecology of *Apodemus* we cannot say which of these factors is the important one limiting its range, or whether either of them is the real reason. For this reason it is fairly useless to make elaborate " laws of distribution " based entirely upon one factor like temperature, as has often been done in the past. It is too crude a method.

12. In field work there is not usually time or opportunity for studying the real limiting factors of species ; and so description of habitats resolves itself into an attempt to record any condition with which a particular species appears to be constantly associated, even if it is a condition which is only correlated with the real limiting factor or factors, and has no significance in the life of the animal. If this is done it is possible for other ecologists to get at least a clear idea of the

exact type of habitat frequented by the animal, and perhaps carry the work still further. There is a species of spider (*Leptyphantes sobrius*) living in Spitsbergen, which illustrates this method. *Leptyphantes sobrius* inhabits a great many types of country, occurring high up on the sides of mountains, or on lowlands, or even on the sea-shore, both in damp and dry places, and it does not appear to be very constantly associated with any particular species of animal or plant. At first sight we might describe its distribution as universal, but careful notes on the habitat of every specimen collected show that there is one feature in its distribution which is absolutely constant. It invariably comes in places where there are rather unstable patches of stones, in stream gullies or on screes, or on the shores of ponds and lakes or of the sea. If there is a patch of soil in one of its typical habitats which has become stabilised through-the agency of plants and is no longer liable to slide away or be washed away during the spring snow-melting, then *Leptyphantes* disappears, or at any rate its numbers will be relatively very small, and other species of spiders (such as *Typhochrestus spetsbergensis*) will take its place. It is worth while in a case like this to be able to define the habitat of the species by some definite factor which is always associated with the animal's distribution, even though, as with *Leptyphantes*, we do not know in the least what the immediate reason for this distribution can be. But it should always be borne in mind that such descriptions of habitat are only a temporary expedient to assist in discovering the *real* habitat, which must be defined in terms of limiting factors. These rough habitat-descriptions are also of great use, since they enable one to make records which assist in the working out of the animal communities of different general habitats.

13. Although we have said that a knowledge of physiology is not always necessary for the solution of ecological problems, it is sometimes of use, especially in cases where the animals with which we are dealing belong to the class which finds its habitat by broadcasting large numbers of young ones, so that only those survive which find a suitable spot. This is true of

many marine animals, and of groups like spiders on land. Vallentin [77b] states that on the Falkland Islands the waters are acid owing to the peaty nature of the soil, and that in winter this acid water runs down on to the shore and rots the shells of the mussels (*Mytilus*) which grow in vast numbers in some places. It is obvious that acidity is a limiting factor to an animal whose shell is made of calcium carbonate. This becomes an important factor with fresh-water molluscs. Waters with a hydrogen-ion concentration of less than 6 seem to be deficient in molluscs. But here again we are in danger of only establishing a correlation, since such molluscs (as also the crayfish) also require a certain minimum salt-content of water in order to build up their shells properly. In these cases the ideal thing is to combine very wide field observations with laboratory experiments in which one factor is varied at a time, the rest being kept constant. This was done by Saunders [76] in the case of the large protozoan *Spirostomum ambiguum*. Studies in the field showed that *Spirostomum* was most active at a pH (hydrogen-ion concentration) of about 7·4, and occurred most commonly in such habitats. This was confirmed by experiments which showed that it died in water of about pH 8 or at a pH below 6, the optimum condition being 7·4. When placed in a gradient of pH it migrated into the region of 7·4 (but for some reason it did this only in the light, regardless, however, of the direction from which the light was coming). This piece of work is an almost ideal example of the way to set about the study of limiting factors : field observation on occurrence and activity followed by experiments on the limits of endurance of the animal, the whole being clinched by putting the animal in a one-factor gradient in order to see what its reactions were.

14. MacGregor [42] has worked out the effects of hydrogen-ion concentration upon the distribution of certain mosquitoes in the south of England. There is one species (*Finlaya geniculata*) which lives as a larva in tree holes containing small amounts of very acid water. The normal pH of these tree-hole waters is well below 4·4. In experiments it was found that the larva develops normally in jars of water whose pH

was below 4·6. But when the pH was allowed to rise, the larvæ died or suspended their development. This was also the case with another tree-hole mosquito (*Anopheles plumbeus*) which flourished in acid and not in alkaline water. On the other hand, two species of mosquitoes (*Anopheles bifurcatus* and *maculipennis*) normally found in alkaline ponds with a pH of about 8·4, died when brought up in acid water of 4·4. The death was partly due to the fact that the water fungus *Saprolegnia*, which is one of the greatest potential enemies of mosquito larvæ, flourishes in acid but not in alkaline water, so that *A. bifurcatus* and *maculipennis* are not normally attacked by it in their alkaline ponds. As soon as the water is made acid *Saprolegnia* can grow and attack the larvæ. *Finlaya* does not normally get attacked by *Saprolegnia*, but when the water is made *slightly* more alkaline than usual the lowered resistance of the larvæ gives the *Saprolegnia* an advantage and causes it to become dangerous to the mosquito.

15. In tropical countries, water-supply is often the chief limiting factor in the ranges of many wild birds and mammals. In the Burmese forests the occurrence of elephants, buffalo, tiger, panther, sambur, barking deer, pig, wild cat, monkeys, etc., is determined in the dry season by the proximity of water-holes, or, in the case of rooting and digging animals, such as pigs and moles, by the softness of the ground, which in turn depends on the water-supply.* In Mesopotamia the black partridge (*Francolinus vulgaris*) is never found more than a hundred yards from water (which it requires for drinking purposes).[2a] The desert quail (*Lophortyx gambeli*) of California is similarly limited by water-supply to areas of scrub by rivers.[2a]

Shelter is sometimes the determining factor in distribution. In North America the round-tailed ground squirrel (*Citellus tereticaudus*), which inhabits deserts, is confined to sandy places which have bushes, since it is diurnal and therefore exposed to the rigours of the hot sun, which it avoids by running from one bush to the next. The big desert kangaroo-rat (*Dipodomys deserti*), which inhabits similar sandy places, is not

* This information was supplied by Capt. C. R. Robbins.

limited by the presence of bushes, since it comes out at night; but it requires deep sand in order that its burrows may be sufficiently cool during the day. The reality of these limiting factors is shown by the fact that both rodents die or are extremely affected if exposed to the sun for any length of time.[2b]

16. In many cases temperature has a direct effect in limiting the activities or in controlling the development of animals. Austin [37] showed that the larvæ of the house fly (*Musca domestica*) died at temperatures of 105° F. or over, and that if piles of horse manure in which there were fly larvæ living were close-packed, the heat generated by decay inside the heap was sufficient to kill off the larvæ except at the edges, and that they · actually migrate away from the hotter parts towards the periphery. Temperature-regulated animals like mammals may also, of course, be severely affected by heat, as shown by the fact that nearly all desert mammals are nocturnal; while the Alaskan fur seal, which is normally adapted to the cold foggy climate of the Pribiloff Islands, suffers a great deal if the temperature rises above 46° to 48° F. At this temperature they show great distress when moving about and fighting, while at 55° to 60° they lie about motionless except that they all fan themselves vigorously with their flippers.[75]

17. Since we do not intend in this book to do more than describe certain principles and methods in ecology, illustrated by examples, we need not enumerate any more examples of the action of physical and chemical factors in limiting the distribution of animals. We have not, however, mentioned any cases of biotic limiting factors, and we shall therefore conclude this chapter by mentioning a few out of the enormous number which exist. It is one of the commonest things in nature to find a herbivorous animal which is attached solely to one plant either for food, or for breeding purposes, or for both. It is hardly necessary to quote examples from amongst insects, since they are so numerous. Such limited choice of plants for food or nesting is not so commonly found among the higher vertebrates, but there are some interesting examples. The beaver (*Castor fiber*) used to occur all over North America from the Colorado River and Northern Florida right up to Alaska

and Labrador, and its northern limit coincides exactly with the northern limit of the aspen (*Populus tremuloides*), which is its favourite and, in those regions, most easily available food. The southern range of the beaver overlaps that of the aspen enormously, and in that part of America the beaver uses, or once used, other trees instead.[74] This example shows the way in which factors are often only limiting to distribution in one part of an animal's range.

18. A species may depend on a plant and an animal, both of which may act as limiting factors in its distribution. The elf owl (*Micropallas whitneyi*), which inhabits parts of the deserts of California and Arizona, is only found in places where the giant cactus (*Cereus giganteus*) grows, since it nests exclusively in that plant. But it is also dependent upon two species of woodpecker (*Centurus uropygialis* and *Colaptes chrysoides mearnsi*) which also nest in the cactus, and whose old nesting-holes are used by the owl. Buxton[20] says : " Less strictly dependent upon the cactus and woodpeckers are a screech owl (*Otus asio gilmani*), a sparrow-hawk (*Falco sparverius*), a flycatcher (*Myiarchus c. cinerascens*), and other birds. None of these are confined to the area inhabited by the Giant Cactus, but they all inhabit that area, and within it they all use old woodpecker holes as nesting sites. Frequently a single trunk of the Giant Cactus contains nests of one or other woodpecker, and also of one of the birds which use the old woodpecker holes. Honey-bees also use these excavations as hives. One must remember that the number of living creatures which eventually depend upon the Giant Cactus includes the scavengers in the birds' nests and bees' nests, the insects, few though they may be, which devour it or frequent its blossoms, and many others. All these organisms depend upon the growth of *Cereus giganteus* for their existence in certain areas."

19. The example that has just been given illustrates the difference between an environmental factor which acts on an animal in the ordinary way, and one which acts as a limiting factor. The woodpeckers are both affected by the presence of the giant cactus, and yet not, as species, absolutely dependent

upon it, whereas the elf owl is. The other birds mentioned are not strictly dependent upon the woodpecker, although its presence affects them enormously, while the elf owl is entirely linked up in its distribution with the woodpecker holes. It further shows that the working out of biotic limiting factors is far more difficult and complex than similar work on physical and chemical factors, since there are such elaborate inter-relations between the different species. We are therefore compelled to study more closely the relations between different animals, and this leads us on to a consideration of animal communities.

CHAPTER V

THE ANIMAL COMMUNITY

"The large fish eat the small fish; the small fish eat the water insects;
the water insects eat plants and mud."
"Large fowl cannot eat small grain."
"One hill cannot shelter two tigers."—CHINESE PROVERBS.

Every animal is (1, 2) closely linked with a number of other animals living round it, and these relations in an animal community are largely food relations. (3) Man himself is in the centre of such an animal community, as is shown by his relations to plague-carrying rats and (4) to malaria or the diseases of his domestic animals, e.g. liver-rot in sheep. (5) The dependence of man upon other animals is best shown when he invades and upsets the animal communities of a new country, e.g. the white man in Hawaii. (6) These interrelations between animals appear fearfully complex at first sight, but are less difficult to study if the following four principles are realised: (7) The first is that of Food-chains and the Food-cycle. Food is one of the most important factors in the life of animals, and in most communities (8) the species are arranged in food-chains which (9) combine to form a whole food-cycle. This is closely bound up with the second principle, (10) the Size of Food. Although animals vary much in size, any one species of animal only eats food between certain limits of size, both lower and (11) upper, which (12) are illustrated by examples of a toad, a fly, and a bird. (13) This principle applied to primitive man, but no longer holds for civilised man, and (14) although there are certain exceptions to it in nature, it is a principle of great importance. (15) The third principle is that of Niches. By a niche, is meant the animal's place in its community, its relations to food and enemies, and to some extent to other factors also. (16), (17), (18), (19), (20) A number of examples of niches can be given, many of which show that the same niche may be filled by entirely different animals in different parts of the world. (21) The fourth idea is that of the Pyramid of Numbers in a community, by which is meant the greater abundance of animals at the base of food-chains, and the comparative scarcity of animals at the end of such chains. (22) Examples of this principle are given, but, as is the case with all work upon animal communities, good data are very scarce at present.

1. If you go out on to the Malvern Hills in July you will find some of the hot limestone pastures on the lower slopes covered with ant-hills made by a little yellow ant (*Acanthomyops flavus*). These are low hummocks about a foot in diameter,

clothed with plants, some of which are different from those of the surrounding pasture. This ant, itself forming highly organised colonies, is the centre of a closely-knit community of other animals. You may find green woodpeckers digging great holes in the ant-hills, in order to secure the ants and their pupæ. If you run up quickly to one of these places, from which a woodpecker has been disturbed, you may find that a robber ant (*Myrmica scabrinodis*) has seized the opportunity to carry off one of the pupæ left behind by the yellow ants in their flight. The latter with unending labour keep building up the hills with new soil, and on this soil there grows a special set of plants. Wild thyme (*Thymus serpyllum*) is particularly common there, and its flowers attract the favourable notice of a red-tailed bumble-bee (*Bombus lapidarius*) which visits them to gather nectar. Another animal visits these ant-hills for a different purpose : rabbits, in common with many other mammals, have the peculiar habit of depositing their dung in particular spots, often on some low hummock or tree-stump. They also use ant-hills for this purpose, and thus provide humus which counteracts to some extent the eroding effects of the woodpeckers. It is interesting now to find that wild thyme is detested by rabbits as a food,[138] which fact perhaps explains its prevalence on the ant-hills. There is a moth (*Pempelia subornatella*) whose larvæ make silken tubes among the roots of wild thyme on such ant-hills ; then there is a great army of hangers-on, guests, and parasites in the nests themselves ; and so the story could be continued indefinitely. But even this slight sketch enables one to get some idea of the complexity of animal interrelations in a small area.

2. One might leave the ants and follow out the effects of the rabbits elsewhere. There are dor-beetles (*Geotrupes*) which dig holes sometimes as much as four feet deep, in which they store pellets of rabbit-dung for their own private use. Rabbits themselves have far-reaching effects upon vegetation, and in many parts of England they are one of the most important factors controlling the nature and direction of ecological succession in plant communities, owing to the fact that they have a special scale of preferences as to food, and eat down

some species more than others. Some of the remarkable results of "rabbit action," on vegetation may be read about in a very interesting book by Farrow.[19] Since rabbits may influence plant communities in this way, it is obvious that they have indirectly a very important influence upon other animals also. Taking another line of investigation, we might follow out the fortunes and activities of the green woodpeckers, to find them preying on the big red and black ant (*Formica rufa*) which builds its nests in woods, and which in turn has a host of other animals linked up with it.

If we turned to the sea, or a fresh-water pond, or the inside of a horse, we should find similar communities of animals, and in every case we should notice that food is the factor which plays the biggest part in their lives, and that it forms the connecting link between members of the communities.

3. In England we do not realise sufficiently vividly that man is surrounded by vast and intricate animal communities, and that his actions often produce on the animals effects which are usually quite unexpected in their nature—that in fact man is only one animal in a large community of other ones. This ignorance is largely to be attributed to town life. It is no exaggeration to say that our relations with the other members of the animal communities to which we belong have had a big influence on the course of history. For instance : the Black Death of the Middle Ages, which killed off more than half the people in Europe, was the disease which we call plague. Plague is carried by rats, which may form a permanent reservoir of the plague bacilli, from which the disease is originally transmitted to human beings by the bites of rat fleas. From this point it may either spread by more rat fleas or else under certain conditions by the breathing of infected air. Plague was still a serious menace to life in the seventeenth century, and finally flared up in the Great Plague of London in 1665, which swept away some hundred thousand people. Men at that time were still quite ignorant of the connection between rats and the spread of the disease, and we even find that orders were given for the destruction of cats and dogs because it was suspected that they were carriers of plague.[149] And there seemed no

reason why plague should not have continued indefinitely to threaten the lives of people in England; but after the end of the seventeenth century it practically disappeared from this country. This disappearance was partly due to the better conditions under which people were living, but there was also another reason. The dying down of the disease coincided with certain interesting events in the rat world. The common rat of Europe had been up to that time the Black or Ship Rat (*R. rattus*), which is a very effective plague-carrier owing to its habit of living in houses in rather close contact with man. Now, in 1727 great hordes of rats belonging to another species, the Brown Rat (*R. norvegicus*), were seen marching westwards into Russia, and swimming across the Volga. This invasion was the prelude to the complete occupation of Europe by brown rats.[87] Furthermore, in most places they have driven out and destroyed the original black rats (which are now chiefly found on ships), and at the same time have adopted habits which do not bring them into such close contact with man as was the case with the black rat. The brown rat went to live chiefly in the sewers which were being installed in some of the European towns as a result of the onrush of civilisation, so that plague cannot so easily be spread in Europe nowadays by the agency of rats. These important historical events among rats have probably contributed a great deal to the cessation of serious plague epidemics in man in Europe, although they are not the only factors which have caused a dying down of the disease. But it is probable that the small outbreak of plague in Suffolk in the year 1910 was prevented from spreading widely owing to the absence of very close contact between man and rats.[71] We have described this example of the rats at some length, since it shows how events of enormous import to man may take place in the animal world, without any one being aware of them.

4. The history of malaria in Great Britain is another example of the way in which we have unintentionally interfered with animals and produced most surprising results. Up to the end of the eighteenth century malaria was rife in the low-lying parts of Scotland and England, as also was liver-rot in

sheep. No one in those days knew the causes or mechanisms of transmission of either of these two diseases; but at about that time very large parts of the country were drained in order to reclaim land for agricultural purposes, and this had the effect of practically wiping out malaria and greatly reducing liver-rot—quite unintentionally! We know now that malaria is caused by a protozoan which is spread to man by certain blood-sucking mosquitoes whose larvæ live in stagnant water, and that the larva of the liver-fluke has to pass through one stage of its life-history in a fresh-water snail (usually *Limnæa truncatula*). The existence of malaria depends on an abundance of mosquitoes, while that of liver-rot is bound up with the distribution and numbers of the snail. With the draining of land both these animals disappeared or became much rarer.[13b]

5. On the whole, however, we have been settled in this country for such a long time that we seem to have struck a fairly level balance with the animals around us; and it is because the mechanism of animal society runs comparatively smoothly that it is hard to remember the number of important ways in which wild animals affect man, as, for instance, in the case of earthworms which carry on such a heavy industry in the soil, or the whole delicately adjusted process of control of the numbers of herbivorous insects. It is interesting therefore to consider the sort of thing that happens when man invades a new country and attempts to exploit its resources, disturbing in the process the balance of nature. Some keen gardener, intent upon making Hawaii even more beautiful than before, introduced a plant called *Lantana camara*, which in its native home of Mexico causes no trouble to anybody. Meanwhile, some one else had also improved the amenities of the place by introducing turtle-doves from China, which, unlike any of the native birds, fed eagerly upon the berries of *Lantana*. The combined effects of the vegetative powers of the plant and the spreading of seeds by the turtle-doves were to make the *Lantana* multiply exceedingly and become a serious pest on the grazing country. Indian mynah birds were also introduced, and they too fed upon *Lantana* berries. After a few

years the birds of both species had increased enormously in numbers. But there is another side to the story. Formerly the grasslands and young sugar-cane plantations had been ravaged yearly by vast numbers of army-worm caterpillars, but the mynahs also fed upon these caterpillars and succeeded to a large extent in keeping them in check, so that the outbreaks became less severe. About this time certain insects were introduced in order to try and check the spread of *Lantana*, and several of these (in particular a species of *Agromyzid* fly) did actually destroy so much seed that the *Lantana* began to decrease. As a result of this, the mynahs also began to decrease in numbers to such an extent that there began to occur again severe outbreaks of army-worm caterpillars. It was then found that when the *Lantana* had been removed in many places, other introduced shrubs came in, some of which are even more difficult to eradicate than the original *Lantana*.[73]

6. It is clear that animals are organised into a complex society, as complex and as fascinating to study as human society. At first sight we might despair of discovering any general principles regulating animal communities. But careful study of simple communities shows that there are several principles which enable us to analyse an animal community into its parts, and in the light of which much of the apparent complication disappears. These principles will be considered under four headings :

    A. Food-chains and the food-cycle.
    B. Size of food.
    C. Niches.
    D. The pyramid of numbers.

Food-chains and the Food-cycle

7. We shall see in a later chapter what a vast number of animals can be found in even a small district. It is natural to ask : " What are they all doing ? " The answer to this is in many cases that they are not doing anything. All cold-blooded animals and a large number of warm-blooded ones

spend an unexpectedly large proportion of their time doing nothing at all, or at any rate, nothing in particular. For instance, Percival [12b] says of the African rhinoceros : "After drinking they play . . . the rhino appears at his best at night and gambols in sheer lightness of heart. I have seen them romping like a lot of overgrown pigs in the neighbourhood of the drinking place."

Animals are not always struggling for existence, but when they do begin, they spend the greater part of their lives eating. Feeding is such a universal and commonplace business that we are inclined to forget its importance. The primary driving force of all animals is the necessity of finding the right kind of food and enough of it. Food is the burning question in animal society, and the whole structure and activities of the community are dependent upon questions of food-supply. We are not concerned here with the various devices employed by animals to enable them to obtain their food, or with the physiological processes which enable them to utilise in their tissues the energy derived from it. It is sufficient to bear in mind that animals have to depend ultimately upon plants for their supplies of energy, since plants alone are able to turn raw sunlight and chemicals into a form edible to animals. Consequently herbivores are the basic class in animal society. Another difference between animals and plants is that while plants are all competing for much the same class of food, animals have the most varied diets, and there is a great divergence in their food habits. The herbivores are usually preyed upon by carnivores, which get the energy of the sunlight at third-hand, and these again may be preyed upon by other carnivores, and so on, until we reach an animal which has no enemies, and which forms, as it were, a terminus on this food-cycle. There are, in fact, chains of animals linked together by food, and all dependent in the long run upon plants. We refer to these as "food-chains," and to all the food-chains in a community as the "food-cycle."

8. Starting from herbivorous animals of various sizes, there are as a rule a number of food-chains radiating outwards, in which the carnivores become larger and larger, while the

parasites are smaller than their hosts. For instance, in a pine wood there are various species of aphids or plant-lice, which suck the juices of the tree, and which are preyed on by spiders. Small birds such as tits and warblers eat all these small animals, and are in turn destroyed by hawks. In an oak wood there are worms in the soil, feeding upon fallen leaves of plants, and themselves eaten by thrushes and blackbirds, which are in turn hunted and eaten by sparrow-hawks. In the same wood there are mice, one of whose staple foods is acorns, and these form the chief food of the tawny owl. In the sea, diatoms form the basic plant food, and there are a number of crustacea (chiefly copepods) which turn these algæ into food which can be eaten by larger animals. Copepods are living winnowing fans, and they form what may be called a "key-industry," in the sea. The term "key-industry" is a useful one, and is used to denote animals which feed upon plants and which are so numerous as to have a very large number of animals dependent upon them. This point is considered again in the section on "Niches."

9. Extremely little work has been done so far on food-cycles, and the number of examples which have been worked out in even the roughest way can be counted on the fingers of one hand. The diagram shown in Fig. 3 shows part of a marine plankton community, which has been studied by Hardy,[102] and which is arranged to show the food-chains leading up to the herring at different times of the latter's life. To complete the picture we should include the dogfish, which attacks the herring itself. Fig. 4 shows the food-cycle on a high arctic island, and is chosen because it is possible in such a place to work out the interrelations of its impoverished fauna fairly completely.

At whatever animal community we look, we find that it is organised in a similar way. Sometimes plants are not the immediate basis of the food-cycle. This is the case with scavengers, and with such associations as the fauna of temporary fresh-water pools and of the abyssal parts of the sea where the immediate basic food is mud and detritus; and the same is true of many parasitic faunas. In all these cases, which

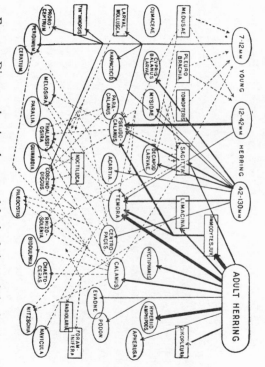

Fig. 3.—Diagram showing the general food relations of the herring to other members of the North Sea plankton community. Note the effect of herring size at different ages upon its food. (From Hardy.[102]

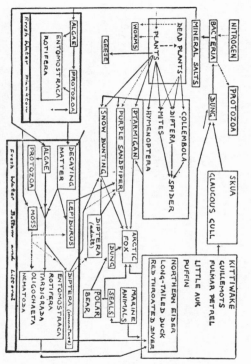

Fig. 4.—Food-cycle among the animals on Bear Island, a barren spot in the arctic zone, south of Spitsbergen. (The dotted lines represent probable food relations not yet proved. The best way to read the diagram is to start at "marine animals" and follow the arrows. (From Summerhayes and Elton.[25])

are peculiar, the food-supply is of course ultimately derived from plants, but owing to the isolation of the animals it is convenient to treat them as a separate community.

Certain animals have succeeded in telescoping the particular food-chain to which they belong. The whale-bone whale manages to collect by means of its sieve-like apparatus enough copepods and pteropods to supply its vast wants, and is not dependent on a series of intermediate species to produce food large enough for it to deal with effectively. This leads us on to a more detailed consideration of the problem of

Size of Food

10. Size has a remarkably great influence on the organisation of animal communities. We have already seen how animals form food-chains in which the species become progressively larger in size or, in the case of parasites, smaller in size. A little consideration will show that size is the main reason underlying the existence of these food-chains, and that it explains many of the phenomena connected with the food-cycle.

There are very definite limits, both upper and lower, to the size of food which a carnivorous animal can eat. It cannot catch and destroy animals *above* a certain size, because it is not strong or skilful enough. In the animal world, fighting weight counts for as much as it does among ourselves, and a small animal can no more tackle a large one successfully than a light-weight boxer can knock out a trained man four stone heavier than himself. This is obvious enough in a broad way: spiders do not catch elephants in their webs, nor do water scorpions prey on geese. Also the structure of an animal often puts limits to the size of food which it can get into its mouth. At the same time a carnivore cannot subsist on animals *below* a certain size, because it becomes impossible at a certain point to catch enough in a given time to supply its needs. If you have ever got lost on the moors and tried to make a square meal off bilberries, you will at once see the force of this reasoning. It depends, however, to a large extent on the numbers of the prey: foxes find it worth while to live entirely on mice in the years when the latter are very abundant, but prey on larger animals like rabbits at other times.

11. It is thus plain that the size of the prey of carnivorous animals is limited in the upward direction by its strength and ability to catch the prey, and in the downward direction by the feasibility of getting enough of the smaller food to satisfy its needs, the latter factor being also strongly influenced by the numbers as well as by the size of its food. The food of every carnivorous animal lies therefore between certain size limits, which depend partly on its own size and partly on other factors. There is an *optimum* size of food which is the one usually eaten, and the limits actually possible are not usually realised in practice. (It is as well to point out that herbivorous animals are not strictly limited by the size of their plant-food, except in special cases such as seed-eating birds, honey-collecting insects etc., owing to the fact that the plants cannot usually run away, or make much resistance to being eaten.) We have very little information as to the exact relative sizes of enemies and their prey, but future work will no doubt show that the relation is fairly regular throughout all animal communities.

12. Three examples will serve to illustrate the part played by size. There lives in the forests round Lake Victoria a kind of toad which is able to adjust its size to the needs of the moment. When attacked by a certain snake the toad swells itself out and becomes puffed up to such an extent that the snake is quite unable to cope with it, and the toad thus achieves its object, unlike the frog in Æsop's fable.[3c] Carpenter[3a] has pointed out another curious case of the importance of size in food. The tsetse fly (*Glossina palpalis*), whose ecology was studied by him in the region of Lake Victoria, can suck the blood of many mammals and birds, in which the size of the blood corpuscles varies from 7 to 18μ, but is unable to suck that of the lungfish, since the corpuscles of the latter (41μ in diameter) are too large to pass up the proboscis of the fly. A third case is that noticed by Vallentin[7b] in the Falkland Islands. He found that the black curlew (*Hæmatopus quoyi*) ate limpets (*Patella ænea*) on the rocks at low tide, but was only

able to dislodge those of moderate size, not usually more than 45 millimetres across.

13. These are three rather curious cases of what is a universal phenomenon. Man is the only animal which can deal with almost any size of food, and even he has only been able to do this during the later part of his history. It appears that the very early ancestors of man must have eaten food of a very limited range of size—such things as shellfish, fruits, mushrooms, and small mammals. Later on, man developed the art of hunting and trapping large animals, and he was thus able to increase the size of his food in the *upward* direction, and this opened up possibilities of obtaining food in greater bulk and variety. After the hunting stage came the agricultural stage, and this consisted essentially in the further development of the use of large animals, now in a domesticated state, and in the invention of means of dealing with foods much *smaller* than had previously been possible, by obtaining great quantities of small seeds in a short time. All other animals except man have their food strictly confined within rather narrow limits of size. The whale-bone whale can feed on tiny crustacea not a thousandth of its bulk, while the killer whale can destroy enormous cuttle-fish ; but it is only man who has the power of eating small, large, and medium-sized foods indiscriminately. This is one of the most important ways in which man has obtained control over his surroundings, and it is pretty clear that if other animals had the same power, there would not be anything like the same variety and specialisation that there is among them, since the elaborate and complex arrangements of the food-cycles of animal communities would automatically disappear. For the very existence of food-chains is due mainly to the fact that any one animal can only live on food of a certain size. Each stage in an ordinary food-chain has the effect of making a smaller food into a larger one, and so making it available to a larger animal. But since there are upper and lower limits to the size of animals, a progressive food-chain cannot contain more than a certain number of links, and usually has less than five.

14. There is another reason why food-chains stop at a

certain point ; this is explained in the section on the Pyramid of Numbers. Leaving aside the question of parasites at present, it may be taken as a fairly general rule that the enemy is larger than the animal upon which it preys. (This idea is contained in the usual meaning of the word "carnivore.") But such is not invariably the case. Fierceness, skill, or some other special adaptation can make up for small size. The arctic skua pursues and terrorises kittiwake gulls and compels them to disgorge their last meal. It does this mainly by naked bluff, since it is, as a matter of fact, rather less in weight than the gull, but is more determined and looks larger owing to a great mass of fluffy feathers. In fact, when we are dealing with the higher animals such as birds, mammals, and the social ants and bees, the psychology of the animals very often plays a large part in determining the relative sizes of enemies and their prey. Two types of behaviour may be noticed. The strength of the prey and therefore its virtual size may be reduced ; this is done by several devices, of which the commonest are poison and fear. Some snakes are able to paralyse and kill by both these methods, and so can cope with larger animals than would otherwise be possible. Stoats are able to paralyse rabbits with fear, and so reduce the speed and strength of the latter. It is owing to this that the stoat can be smaller than its prey. The fox, which does not possess this power of paralysing animals with fear, is considerably larger than the rabbit. The second point is that animals are able to increase their own effective size by flock tactics. Killer whales in the Antarctic seas have been seen to unite in parties of three or four in order to break up the thick ice upon which seals, their prey, are sleeping.[150] Wolves are another example. Most wolves are about half the linear size of the deer which they hunt, but by uniting in packs they become as formidable as one very large animal. The Tibetan wolf, which eats small gazelles, etc., hunts singly or in twos and threes.[43a] On the other hand, herbivores often band together in flocks in order to increase their own powers of defence. This usually means increased strength, but other factors come in too. Ants have achieved what is perhaps the most successful solution of the size problem, since they form organised colonies whose size is entirely fluid according to circumstances. Schweitzer[88] noted a column of driver ants in Angola march past for thirty-six hours. They are able by the mass action of their terrible battalions to destroy animals many times their own size (*e.g.* whole litters of the hunting dog[121]), and at the same time can carry the smallest of foods.

It must be remembered, therefore, that the idea of food-chains of animals of progressively larger size is only true in a general way, and that there are a number of exceptions. Having considered the far-reaching effects of size on the organisation of animal communities, we are now in a position to consider the subject of

Niches

15. It should be pretty clear by now that although the actual species of animals are different in different habitats, the ground plan of every animal community is much the same. In every community we should find herbivorous and carnivorous and scavenging animals. We can go further than this, however: in every kind of wood in England we should find some species of aphid, preyed upon by some species of lady-bird. Many of the latter live exclusively on aphids. That is why they make such good controllers of aphid plagues in orchards. When they have eaten all the pest insects they just die of starvation, instead of turning their attention to some other species of animal, as so many carnivores do under similar circumstances. There are many animals which have equally well-defined food habits. A fox carries on the very definite business of killing and eating rabbits and mice and some kinds of birds. The beetles of the genus *Stenus* pursue and catch springtails (*Collembola*) by means of their extensile tongues. Lions feed on large ungulates—in many places almost entirely zebras. Instances could be multiplied indefinitely. It is therefore convenient to have some term to describe the status of an animal in its community, to indicate what it is *doing* and not merely what it looks like, and the term used is " niche." Animals have all manner of external factors acting upon them—

chemical, physical, and biotic—and the "niche" of an animal means its place in the biotic environment, *its relations to food and enemies*. The ecologist should cultivate the habit of looking at animals from this point of view as well as from the ordinary standpoints of appearance, names, affinities, and past history. When an ecologist says "there goes a badger" he should include in his thoughts some definite idea of the animal's place in the community to which it belongs, just as if he had said "there goes the vicar."

16. The niche of an animal can be defined to a large extent by its size and food habits. We have already referred to the various key-industry animals which exist, and we have used the term to denote herbivorous animals which are sufficiently numerous to support a series of carnivores. There is in every typical community a series of herbivores ranging from small ones (*e.g.* aphids) to large ones (*e.g.* deer). Within the herbivores of any one size there may be further differentiation according to food habits. Special niches are more easily distinguished among carnivores, and some instances have already been given.

The importance of studying niches is partly that it enables us to see how very different animal communities may resemble each other in the essentials of organisation. For instance, there is the niche which is filled by birds of prey which eat small mammals such as shrews and mice. In an oak wood this niche is filled by tawny owls, while in the open grassland it is occupied by kestrels. The existence of this carnivore niche is dependent on the further fact that mice form a definite herbivore niche in many different associations, although the actual species of mice may be quite different. Or we might take as a niche all the carnivores which prey upon small mammals, and distinguish them from those which prey upon insects. When we do this it is immediately seen that the niches about which we have been speaking are only smaller subdivisions of the old conceptions of carnivore, herbivore, insectivore, etc., and that we are only attempting to give more accurate and detailed definitions of the food habits of animals.

17. There is often an extraordinarily close parallelism

between niches in widely separated communities. In the arctic regions we find the arctic fox which, among other things, subsists upon the eggs of guillemots, while in winter it relies partly on the remains of seals killed by polar bears. Turning to tropical Africa, we find that the spotted hyæna destroys large numbers of ostrich eggs, and also lives largely upon the remains of zebras killed by lions.[12a] The arctic fox and the hyæna thus occupy the same two niches—the former seasonally, and the latter all the time. Another instance is the similarity between the sand-martins, which one may see in early summer in a place like the Thames valley, hawking for insects over the river, and the bee-eaters in the upper part of the White Nile, which have precisely similar habits. Both have the same rather distinct food habits, and both, in addition, make their nests in the sides of sand cliffs forming the edge of the river valleys in which they live. (Abel Chapman[85a] says of the bee-eaters that " the whole cliff-face appeared aflame with the masses of these encarmined creatures.") These examples illustrate the tendency which exists for animals in widely separated parts of the world to drift into similar occupations, and it is seen also that it is convenient sometimes to include other factors than food alone when describing the niche of any animal. Of course, a great many animals do not have simple food habits and do not confine themselves religiously to one kind of food. But in even these animals there is usually some regular rhythm in their food habits, or some regularity in their diverse foods. As can be said of every other problem connected with animal communities, very little deliberate work has been done on the subject, although much information can be found in a scattered form, and only awaits careful coordination in order to yield a rich crop of ideas. The various books and journals of ornithology and entomology are like a row of beehives containing an immense amount of valuable honey, which has been stored up in separate cells by the bees that made it. The advantage, and at the same time the difficulty, of ecological work is that it attempts to provide conceptions which can link up into some complete scheme the colossal store of facts about natural history which has accumulated up to date in this rather hap-hazard manner. This applies with particular force to facts about the food habits of animals. Until more organised information about the subject is available, it is only possible to give a few instances of some of the more clear-cut niches which happen to have been worked out.

18. One of the biggest niches is that occupied by small sap-suckers, of which one of the biggest groups is that of the plant-lice or aphids. The animals preying upon aphids form a rather distinct niche also. Of these the most important are the coccinellid beetles known as ladybirds, together with the larvae of syrphid flies (cf. Fig. 5) and of lacewings. The niche

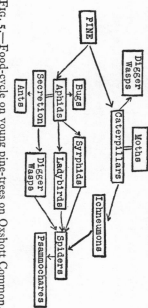

FIG. 5.—Food-cycle on young pine-trees on Oxshott Common. (From Richards.[18])

in the sea and in fresh water which is analogous to that of aphids on land is filled by copepods, which are mainly diatom-eaters. This niche occurs all over the world, and has a number of well-defined carnivore niches associated with it. If we take a group of animals like the herbivorous grass-eating mammals, we find that they can be divided into smaller niches according to the size of the animals. There is the mouse niche, filled by various species in different parts of the world ; the rabbit niche, of larger size, filled by rabbits and hares in the pala-arctic region and in North America, by the agouti and viscacha in South America, by wallabies in Australia, and by animals like the hyrax, the springbuck, and the mouse deer [56] in Africa. In the same way it can be shown that there is a special niche of carnivorous snakes which prey upon other snakes—a niche which is filled by different species in different countries. In

South America there is the mussarama, a large snake four or five feet in length, which is not itself poisonous, but preys exclusively upon other snakes, many of which are poisonous, being itself immune to the venoms of lachesis and rattlesnake, but not to colubrine poisons. In the United States the niche is filled by the king-snake which has similar habits, while in India there is a snake called the hamadryad which preys upon other (in this case non-poisonous) snakes.[86a]

19. Another widespread niche among animals is that occupied by species which pick ticks off other animals. For instance, the African tick-bird feeds entirely upon the ticks which live upon the skin of ungulates, and is so closely dependent upon its mammalian "host" that it makes its nest of the latter's hair (e.g. of the hartebeest).[12f] In England, starlings can often be seen performing the same office for sheep and deer. A similar niche is occupied on the Galapagos Islands by a species of scarlet land-crab, which has been observed picking ticks off the skin of the great aquatic lizards (Ambly-rhynchus).[36c] Another niche, rather analogous to the last one, is that occupied by various species of birds, which follow herds of large mammals in order to catch the insects which are disturbed by the feet of the animals. Chapman [85d] saw elephants in the Sudan being followed by kites and grey herons; Percival [12g] says that the buff-backed egret follows elephants and buffalo in Kenya for the same purpose; in Paraguay [86e] there are the Aru blackbirds which feed upon insects disturbed by the feet of cattle; while in England wagtails attend cattle and sheep in the same way.

20. There is a definite niche which is usually filled by earth-worms in the soil, the species of worm differing in different parts of the world. But on coral islands their place may be largely taken by land-crabs. Wood-Jones [107a] states that on Cocos-Keeling Island, coconut husks are one of the most important sources of humus in the soil, and in the rotting husks land-crabs (chiefly of the genus Cardiosoma) make burrows and do the same work that earthworms do in our own country. (There are as a matter of fact earthworms as well on these islands.) On the coral reefs which cover such a large

part of the coast in tropical regions, there is a definite niche filled by animals which browse upon the corals, just as herbivorous mammals browse upon vegetation on land. There are enormous numbers of holothurians or sea-cucumbers which feed entirely in this way. Darwin [30] gives a very good description of this niche. Speaking also of Cocos-Keeling Island, he says :

"The number of species of Holothuria, and of the individuals which swarm on every part of these coral-reefs, is extraordinarily great ; and many ship-loads are annually freighted, as is well known, for China with the trepang, which is a species of this genus. The amount of coral yearly consumed, and ground down into the finest mud, by these several creatures, and probably by many other kinds, must be immense. These facts are, however, of more importance in another point of view, as showing us that there are living checks to the growth of coral-reefs, and that the almost universal law of ' consume and be consumed,' holds good even with the polypifers forming those massive bulwarks, which are able to withstand the force of the open ocean."

"This passage, besides showing that the coral-eating niche has a geological significance, illustrates the wide grasp of ecological principles possessed by Darwin, a fact which continually strikes the reader of his works. We have now said enough to show what is meant by an ecological niche, and how the study of these niches helps us to see the fundamental similarity between many animal communities which may appear very different superficially. The niche of an animal may to some extent be defined by its numbers. This leads us on to the last subject of this chapter,

The Pyramid of Numbers

21. " One hill cannot shelter two tigers." In other and less interesting words, many carnivorous animals, especially at or near the end of a food-chain, have some system of territories, whereby it is arranged that each individual, or pair, or family, has an area of country sufficiently large to supply its food requirements. Hawks divide up the country in this way,

and Eliot Howard's work [9] has shown that similar territory systems play a very important part in the lives of warblers. We can approach the matter also from this point of view : the smaller an animal the commoner it is on the whole. This is familiar enough as a general fact. If you are studying the fauna of an oak wood in summer, you will find vast numbers of small herbivorous insects like aphids, a large number of spiders and carnivorous ground beetles, a fair number of small warblers, and only one or two hawks. Similarly in a small pond, the numbers of protozoa may run into millions, those of *Daphnia* and *Cyclops* into hundreds of thousands, while there will be far fewer beetle larvæ, and only a very few small fish. To put the matter more definitely, the animals at the base of a food-chain are relatively abundant, while those at the end are relatively few in numbers, and there is a progressive decrease in between the two extremes. The reason for this fact is simple enough. The small herbivorous animals which form the key-industries in the community are able to increase at a very high rate (chiefly by virtue of their small size), and are therefore able to provide a large margin of numbers over and above that which would be necessary to maintain their population in the absence of enemies. This margin supports a set of carnivores, which are larger in size and fewer in numbers. These carnivores in turn can only provide a still smaller margin, owing to their large size which makes them increase more slowly, and to their smaller numbers. Finally, a point is reached at which we find a carnivore (*e.g.* the lynx or the peregrine falcon) whose numbers are so small that it cannot support any further stage in the food-chain. There is obviously a lower limit in the density of numbers of its food at which it ceases to be worth while for a carnivore to eat that food, owing to the labour and time that is involved in the process. It is because of these number relations that carnivores tend to be much more wide-ranging and less strictly confined to one habitat than herbivores.

22. This arrangement of numbers in the community, the relative decrease in numbers at each stage in a food-chain, is characteristically found in animal communities all over the

world, and to it we have applied the term "pyramid of numbers." It results, as we have seen, from the two facts (a) that smaller animals are preyed upon usually by larger animals, and (b) that small animals can increase faster than large ones, and so are able to support the latter.

The general existence of this pyramid in numbers hardly requires proving, since it is a matter of common observation in the field. Actual figures for the relative numbers of different stages in a food-chain are very hard to obtain in the present state of our knowledge. But three examples will help to crystallise the idea of this "pyramid." Birge and Juday [92] have calculated that the material which can be used as food by the plankton rotifers and crustacea of Lake Mendota in North America weighs twelve to eighteen times as much as they do. (The fish which eat the crustacea would weigh still less.) Again, Mawson [93] estimated that one pair of skuas (*Megalestris*) on Haswell I. in the Antarctic regions, required about fifty to one hundred Adelie penguins to keep them supplied with food (in the form of eggs and young of the penguins); while Percival [124] states that one lion will kill some fifty zebras per year, which gives us some idea of the large numbers of such a slow-breeding animal as the zebra which are required to produce this extra margin of numbers.

CHAPTER VI

PARASITES

As opposed to carnivores, parasites are (1) much smaller than the hosts upon which they prey, but (2) feed in essentially the same way as carnivores, the chief difference being that the latter live on capital and the former on income of food ; but (3) a complete graded series can be traced between typical parasites and typical carnivores, both among animals which eat other ones, and (4) among those animals which live by robbing other animals of their food. (5) We can therefore apply the same principles to parasites as were applied to carnivores in the last chapter, making certain alterations as a result of the different size-relations of the two classes of animals. (6) The food-cycle acts as an important means of dispersal for internal parasites, so that they often have two, or (7) more, hosts during their life-histories, and (8) this means of dispersal is also used to some extent by external parasites. Animals which suck blood or plant-juices often play an important part in such life-histories. (9) In food-chains formed by parasites, the animals at each stage become smaller, and at the end of a chain are so small that bacteria are often found to become important organisms at that point. (10) The parasitic hymenoptera occupy a rather special position in the food-cycle. (11) When parasites and carnivores are both included in the same scheme of food-cycles, the latter become very complex, as is (12) shown by an example ; but (13) in practice, a number of parasites can be considered as forming part of their host, as far as food is concerned, although when numbers are being studied the parasites must be treated separately.

1. If you make a list of the carnivorous enemies and of the parasites of any species of animal, you will see (although they are so obvious that they easily escape notice unless pointed out) certain curious facts about the sizes of the two classes of animals relative to their prey or host. For instance, a frog would have for enemies such animals as otters, herons, or pike, which would be anything up to fifty times the size of the frog, while the parasites would be animals like flat-worms, nematodes, or protozoa, which would be five hundred or five thousand times smaller than the frog. The same thing applies to any other animal. A mouse is preyed on by

hawks, owls, foxes, and weasels on the one hand, and by lice, fleas, ticks, mites, tapeworms, and protozoa on the other. Small oligochæte worms in the arctic soil are eaten by purple sandpipers and parasitised by protozoa, while earthworms in England are eaten by moles and shrews and thrushes and toads, and parasitised by protozoa and nematodes. In fact, most animals have a set of carnivorous animals much larger than themselves, and a set of parasitic enemies much smaller than themselves, and usually there are very few enemies of intermediate size. In all these cases we are, of course, speaking of the size relative to their prey or host. There are perfectly good reasons for this size-distribution. Most carnivores are able to overpower and eat their prey by virtue of their larger size and greater strength (with exceptions noted in the last chapter). On the other hand, parasites must necessarily be much smaller than their hosts, since their existence depends, either temporarily or permanently, upon the survival of the host, and for this reason the parasite cannot exceed a certain size without harming its host too much.

2. It is very important to realise quite clearly that most parasites are in their feeding habits doing essentially the same thing as carnivores, except that while the carnivore destroys its prey, the parasite does not do so, or at any rate does not do so immediately or completely. A parasite's existence is usually an elaborate compromise between extracting sufficient nourishment to maintain and propagate itself, and not impairing too much the vitality, or reducing the numbers of its host, which is providing it with a home and a free ride. In consequence of this compromise, a parasite usually destroys only small portions of its host at a time, portions which can often be replaced fairly quickly by regeneration of the tissues attacked. Or it may exploit the energies of its host in more subtle ways, as when it subsists on the food which the host has collected with great expenditure of time and energy. The difference between the methods of a carnivore and a parasite is simply the difference between living upon capital and upon income; between the habits of the beaver, which cuts down a whole tree a hundred years old, and the bark-beetle, which levies a

daily toll from the tissues of the tree; between the burglar and the blackmailer. The general result is the same, although the methods employed are different.

3. Although the relative sizes of carnivores and parasites are so markedly different in the vast majority of cases, it is really rather difficult to draw the line in all cases between the two classes of animals, since there are a number of species which combine to some extent the characteristics of both types. It is possible and interesting to trace a complete series between the two extremes. A hookworm lodged in the intestine of a mammal is a typical parasite, destroying the tissues of its host, and, in the adult stage, entirely confined to it. A louse is also clearly a typical parasite, although it can walk about from one host to another. Fleas, however, are less constant in their attendance upon their host, since they are able to live for some days at a time without feeding, and may be found walking about at large, in the open or in nests of their host. When we come to blood-sucking flies it is quite difficult to know whether to class them as parasitic or carnivorous. Their habits and size may be same as those of fleas, and the time which is spent with the host may be no less than that of the flea, but they are less closely attached to any one host. This time varies considerably : a complete series could be traced between species which spend a great deal of their time in the company of their host-animal (*e.g.* the horse-fly (*Hippobosca equina*) which lives close to its host), and others which lead highly independent lives (*e.g.* the Tabanids, which are generally active fliers and sit about waiting for some animal to pass).

4. We can find a similar graded series amongst animals which live by robbing other animals of their food. The adults of certain *Filaria* worms live embedded in the muscles of their hosts. The larvæ of these worms live in the blood or the lymph-vessels of the host. Then there are tapeworms, which inhabit the intestine and absorb food which has previously been made soluble by the digestive juices of the host. Other animals (*e.g.* nematode worms), often living in the same place as the tapeworms, exist by eating solid food particles, which

are either undigested or only partly digested. The next stage in this series is the crocodile bird, which sits inside the open mouth of the crocodile and picks bits of food from amongst the teeth of its "host." The arctic skua employs similar but more drastic methods, when it chases kittiwake gulls and terrorises them into yielding up their last meal, which is skil-fully caught by the skua before reaching the water below. Another method employed by some animals is to take small bits out of their food-animal, without actually destroying it, just after the manner of the hookworm or the filaria. For instance, Lortet [53] says : "The fishes of the lake of Tiberius [Tiberias], very good to eat, serve as a pasturage for the myriads of crested grebes (*Podiceps cristatus*) and of pelicans. Frequently the grebes snatch at the eyes of the chromids, and with one stroke of their long sharp beaks lift out as cleverly as would a skilful surgeon the two eyeballs and the intro-orbital partition. These unhappy fish, now blind, of which we have taken numerous examples, have thus the entire face perforated by a bloody canal which cicatrises rapidly. It is only the larger individuals who are thus operated on by the grebes, for, not being able to avail themselves of the entire fish, these voracious birds take the precaution to snatch only the morsel of their choice." Put less poetically, this means that the bird is able to carry on the ordinary carnivorous method of destroying the whole of its prey as long as the latter is below a certain size, but when it grows above this limit new methods are adopted, which closely resemble those of a true parasite. A further stage in the series which we have been tracing is the arctic fox, which, although an ordinary carnivorous animal in summer, when it eats birds and lemmings, often travels out on to the frozen sea-ice in winter and there accompanies the polar bear and subsists on the remains of seals killed by the bear, and upon the dung of the latter. The bear does all the work, and the fox gets a share of the proceeds. From this point it is only a short distance to a true carnivore, like the polar bear itself, which is, after all, only living by exploiting the energies of the seal.

5. To imagine that parasites are unique in exploiting the activities and food-products of their hosts is to take a very

limited view of natural history. It is common to find parasites referred to as if they were in some way more morally oblique in their habits than other animals, as if they were taking some unfair and mean advantage of their hosts. If we once start working out such " responsibilities " we find that the whole animal kingdom lives on the spare energy of other species or upon plants, while the latter depend upon the radiant energy of the sun. If parasites are to occupy a special place in this scheme we must, to be consistent, accuse cows of petty larceny against grass, and cactuses of cruelty to the sun. Once we take a broad view of animal interrelations it becomes quite clear that it is best to treat parasites as being essentially the same as carnivores, except in their smaller size, which enables them to live on their host. In other words, the resemblances between the two classes of animals are more important than the differences.

6. We will now turn to a consideration of the place of parasites in the food-cycle of any animal community, and the ways in which the food-cycle affects them. One of the greatest questions which has to be solved by many parasites is what to do when their host dies, as it is bound to do sooner or later. This applies with especial force to internal parasites like flatworms, which have become so specialised to a life of passive absorption in the dark that they are unable to take any active steps to deal with the situation created by the death of their host. It is here that the food-cycle comes in and plays an important part. Probably the commonest death for many animals is to be eaten by something else, and as a result we find that a great many parasites pass automatically with the prey into the body of its enemy, and are then able in some way to occupy the new host. Let us take the case of a tapeworm which lives as a young larva or bladderworm in the muscles of a rabbit. When the rabbit is eaten by some enemy, say a fox, some of the bladderworms pass unharmed into the intestine of the fox, and there continue their development and grow up into adult tapeworms ; and, in this way, the problem created by the death of the first host is solved. But foxes being also mortal, the tapeworm has to get back again into the rabbit before the

fox dies, and this is also brought about by the food-cycle. For the tapeworm produces vast numbers of eggs which pass out with the excretory products of the fox ; some of these eggs contaminate the vegetation which the rabbit is eating, or in some other way get in with its food, and are then able ultimately to grow up into more bladderworms in the body of the rabbit. The diagram in Fig. 6 shows the way in which the food-cycle

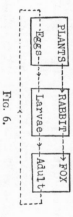

FIG. 6.

acts as a means of conveyance for the parasites, throughout its life-cycle.

7. This was a fairly simple case. There is always this tendency for parasites to get transferred from one stage in a food-chain to the next, like passengers on a railway. Many parasites get out at the first station—in other words, they have a direct life-history, with no alternate host. An example of this is a tapeworm (*Hymenolepis*) which occurs in mice, and which has the larval and adult stages in the same host, although in different parts of the body.[54] Or the parasite may get out at the second station, like the rabbit tapeworm described above. Or again, it may travel as far as a third host. The broad tapeworm (*Diphyllobothrium latum*), which occurs occasionally in man, causing severe anaemia, gets into the gut of some small fresh-water copepod (e.g. *Cyclops strenuus* or *Diaptomus*

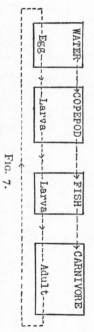

FIG. 7.

gracilis) with the food of the latter, is then eaten by fish in which it exists in the form of bladderworm cysts, and is finally eaten by some carnivore or by man.[54] The diagram (Fig. 7) sums up this cycle.

Perhaps the most striking examples of the way in which parasites may pass along food-chains are afforded by the worms of the genus *Echinorhynchus*, which in some cases go on being transmitted from one host to another until they end up in the animal which forms the end of a whole chain, and so can get no farther, and indeed probably never get back again to their first host. They are like passengers who forget to get out at the right station and travel by mistake on to the terminus.

8. There is often a similar tendency amongst ectoparasites to become transferred from their first host to its enemy. This is well known to be the case amongst some species of fleas. *Ceratophyllus sciurorum*, which lives commonly upon the squirrel and dormouse, occurs also occasionally on the pine-marten, weasel, and stoat, while *C. walkeri* comes commonly on the bank-vole, weasel, and stoat.[55]

All the cases in which parasites are transmitted from one host to another by blood-sucking insects are also examples of the rôle played by the food-cycle in the lives of parasites. There are parallel cases of plant parasites being carried from plant to plant by insects. The protozoan *Phytomonas*, which is found very widely in the tissues of various species of spurge (*Euphorbia*), mainly in warmer countries, is carried from one plant to another, in some places at least, by insects. It has been shown that *Phytomonas davidi*, which occurs in *Euphorbias* in Portugal, is carried from plant to plant by a bug, *Stenocephalus agilis*, which lives upon the juices of the spurge, and in which the *Phytomonas* has a definite life-history stage.[89a]

Since the food-cycle is so important in determining many of the possible modes of transference of parasites from one host to another, it is plain that biological surveys carried out along food-cycle lines would be of great value in narrowing the field of inquiry when the life-history of any particular parasite is being studied.

9. Having shown the relation which exists between parasites and food-chains formed by herbivorous and carnivorous animals, we may now turn to a consideration of food-chains among parasites themselves. Just as the carnivores in a food-chain usually become progressively larger and larger,

so of course the animals in a parasite food-chain become gradually smaller and smaller. And just as the carnivores become fewer in numbers, so do the parasites become usually more numerous. Let us take some examples. Fleas are parasitic upon birds and mammals, and many fleas in turn are parasitised by protozoa of the genus *Leptomonas*.[89b] One squirrel might support a hundred fleas or more, and each flea might support thousands or hundreds of thousands of *Leptomonas*. The pyramid of numbers in such a case is inverted. Many other examples could be given. Egyptian cattle have ticks (*Hyalomma*), which in turn carry inside them a protozoan (*Crithidia hyalommœ*).[89c] Apparently there are never very many stages in such food-chains of parasites. The reason for this is that the largest parasite is not very big, and any hyperparasite living on or in this must be very much smaller still, so that the fifth or sixth stage in the chain would be something about the size of a molecule of protein! Actually, bacteria, although they are not animals, may be conveniently included in parasite food-cycles, and in many cases they form the last link in the chain. The plague bacillus lives in the flea of the rat in warm countries (also in the rat itself), the flea lives on the rat, and the rat lives to some extent "parasitically" on mankind. Recent work upon bacteriophages seems to show that there are sometimes still smaller organisms (or something with the power of multiplication) which have a controlling effect on the numbers of the bacteria themselves. So possibly the last link in some parasite food-chains may be formed by such bacteriophages.

10. There has been a good deal of discussion by entomologists as to the exact position of the Parasitic Hymenoptera with respect to parasitism. These insects lay their eggs in the eggs, larvæ, or later stages of various other insects (and in some other animals too), and these eggs grow into larvæ which live as true parasites in the bodies of their hosts, which they eventually destroy, the hymenoptera finally emerging in the form of free-living insects. An animal like an ichneumon, therefore, combines the characteristics of a parasite (in its larval stage) and of a free-living animal (in its adult stage),

the latter being carnivorous, or herbivorous, or in some cases not feeding at all. For this reason the Parasitic Hymenoptera have often been referred to as "parasitoids." There are numerous other kinds of animals which have alternate free-living and parasitic stages in their life-history, but not many of them in which the adult is free-living as it is in these hymenoptera. It is usually the larva which undertakes the task of finding a new host, where this is done by active migration.

As Richards [18b] has pointed out, insects like ichneumons, braconids, and chalcids, do not have directly a very big effect upon the food-cycle in a community, since they are merely turning the tissues of their host into hymenopteran tissue, and very often the enemies of the host eat the adult hymeno-ptera as readily as they would eat the original host if it had survived. Of course it does make a certain amount of differ-ence ; for one thing, there is naturally a great loss of energy in the process of turning host into parasite, and therefore the activity of the hymenoptera reduces the available food-supply to some extent.

II. It has been necessary in this chapter, as well as in the last, to speak continually of food-chains as if they commonly consisted of simple series of species, without taking into account any of the complications found in actual practice. This is of course far from the truth, because the food-relations of animals are extremely complicated and form a very closely and intricately woven fabric—so elaborate that it is usually quite impossible to predict the precise effects of twitching one thread in the fabric. Simple treatment of the subject makes it possible to obtain a glimmering of the principles which underlie the superficial complication, although it must be clearly recognised that we know at present remarkably little about the whole matter. One of the most important of these principles is that the sizes, and in particular the relative sizes, of the various animals which live together in a com-munity, play a great part in their lives, and partly determine the effects which the various species will have upon one another. One important effect we have mentioned—that the food-cycle provides a means of dispersal for parasites, a means which is very commonly employed by them. The most striking effects, however, are concerned with the numbers of animals; but this subject must be reserved for another chapter.

12. When the food-relations of parasites and carnivores to other species are combined into a common food-cycle scheme, the amazingly complex nature of animal interrelations is seen. The diagram in Fig. 8 is intended to give some idea of the sort of thing which is met with; but it is to be understood that the diagram is only illustrative and does not claim to have anything but a general basis of truth, i.e. it does not

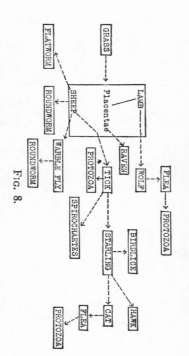

Fig. 8.

represent any particular community in real life, and "all the characters in the story are entirely fictitious." If there were plenty of real examples to choose from, it would be much better and more satisfactory; but there are not. It is this very fact—the lack of properly worked-out examples—which makes it important to try and point out the type of problem which requires solution. The diagram illustrates several points. For one thing it shows that the food-chains do not always go on from carnivore to carnivore, or parasite to parasite. In this case the tick which is parasitic upon the sheep is eaten by the starling, which is a carnivore, the latter in turn by an animal larger than itself (the cat), and the cat in turn by fleas, and the fleas by protozoa. In this way, when parasites and carnivores are considered together it is seen that there may be a great many more stages in the history of the food than when parasites alone or carnivores alone are considered. In this example, in which mutton is converted, amongst other things, into flea-protozoa, there are six stages involved, whereas the simple carnivore chain from the sheep to the wolf has only two stages, and the simple parasite chains only two or three at the most. The reason for the greater number of stages possible in the former case is simply that the size-relationships of the animals make it possible for the chain to continue longer without reaching either the upper or lower size limits of animals.

13. A great many ectoparasites have no very important direct effects upon the food-cycle in general, since they are eaten either by their hosts (*e.g.* birds and bird-lice) or else they are eaten together with their host by the enemy of the latter (*e.g.* copepods containing worm-larvæ by fish). In such cases the parasite and its host act as one unit for food purposes. (Of course the circulation of the parasite in this way may ultimately have very important effects upon the numbers of animals in the community.) It is for this reason that parasites can very often be ignored in practice, when one is making out the first rough scheme of food-relations in an animal community, although the control of numbers can only be understood by bringing in the parasites too. For instance, in the arctic tundra food-cycle shown in Fig. 4, on p. 58, the parasites do not play a very important part, and probably it is the exception rather than the rule for parasites to form a large independent food-supply for any other animal. There are, however, a number of ectoparasites (*e.g.* ticks) occurring upon mammals, which form an important article of diet for certain species of birds. There is one rather interesting instance of this sort of thing, recorded by Wilkins [34] from the antarctic regions. On Elephant Island there is in summer a colony of nesting Gentoo Penguins (*Pygoscelis papua*) and haunting these colonies are a certain number of birds called Paddies (*Chionis alba*). The Paddies live largely upon parasitic nematode worms which pass out from the intestine of the penguins with their excreta. In

winter the Paddies become very thin, owing to the absence
of the penguins upon which they depend in summer.

These instances show that it is not possible to neglect the
existence of parasites as a food-supply for other animals, but
that they do not usually act as such except in the sense of
being eaten at the same time as their host, by the latter's
enemy.

CHAPTER VII

TIME AND ANIMAL COMMUNITIES

Many of the animals in a community (1) never meet owing to the fact that they become active at different times. This is because (2) the environment is subject to a number of rhythmical changes which (3) result in corresponding variations in the nature of the animal communities at different times. (4) There is the day and night rhythm which affects both free-living animals and (5) some parasites. (6) This rhythm may be of practical importance, e.g. from its influence upon blood-sucking insects and (7) is most strongly marked in deserts, but (8) there is really very little detailed information about day and night communities. (9) Some of the changes in the fauna are caused by migration, as in the vertical strata of a wood or in the plankton. (10) There is not always a very sharp limit between day and night communities. (11) In polar regions there is no night fauna, while (12) in the tropics the latter is very rich. (13) Other rhythms are those of the tides and (14) of weather (caused by the passage of depressions) which (15) produce variations in the composition of the active animal communities, e.g. those of dry and (16) wet conditions, which (17) may override the day and night rhythm. (18) Weather changes also have important effects on blood-sucking insects. (19) Then there is the annual cycle of the seasons which (20) has a particular interest from its relation to bird migration and (21) to changes both in the food-habits of animals and (22) in the particular species occupying any one niche. (23) These rhythmical changes in communities enormously increase the difficulty of studying the latter completely, so that (24) it is advisable to choose extremely simple ones in order to work out the principles governing animal communities in general. (25) Larger pulsations in climate chiefly affect the numbers of animals, and so we are led on to the next chapter.

1. ONE of the commonest rodents of the South African veld is the gerbille, springhaasrot, or rooiwitpens (*Taterona lobengula*), which lives sociably in warrens in sandy country where there is plenty of sweet grass and bulbs to eat. Quite often it makes its network of burrows in places already occupied by two species of carnivorous animals, the yellow mongoose (*Cynictis pencillata*) and the suricat (*Suricator suricator*). But although the rodents and the carnivores live in close contact with one another, actually using to some extent the same system

of underground runways, they do not usually clash in any way in their activities ; for while the gerbilles come out exclusively at night, leaving their burrows after sunset and returning always before dawn, the mongooses and suricats feed only during the day, and retire to earth at night.[70] It appears that a mongoose never attacks a gerbille under ordinary circumstances because the two creatures do not usually meet : they have different hours for business. It is only when the gerbilles are smitten by an epidemic of some disease like plague, at which times they wander out of their holes in the daytime, that they are attacked and eaten by the mongoose. This last fact has a practical importance, since the South African plague investigator is able by examining the excreta of the mongoose to find out with tolerable certainty whether the gerbilles have been dying of plague, a fact which is rather difficult to establish easily in any other way. If the mongoose excreta contain gerbille fur, then there is strong evidence of epidemic amongst the latter.[70]

In this case the alternation of day and night has the effect of separating almost completely two animals which live in the same place, and although the phenomenon and its results happen to have an important practical bearing, it is only one example among thousands which might be given, all of which go to show that the phenomenon is of general occurrence in nearly all animal communities.

2. The environment even in the same place is always changing rhythmically and more or less violently ; some of these changes being regular, like the alternation of day and night, of high and low tides, or the annual succession of the seasons, while others are more irregular, like the fluctuations in weather from day to day and week to week. These changes all leave a corresponding impress upon the arrangement and composition of animal communities. Just as animals tend to become specialised for life in certain places, so also most of them are active only at certain times. There are various ways of meeting the onset of unfavourable conditions. If the latter last only for a short time, the animal may merely retire to some hiding-place or become inactive wherever it happens to be at the moment. Every one must have noticed

the extraordinary effect upon insects when a passing cloud covers the sun. The drop in temperature slows down their movements or actually stops them altogether. The return of the sun starts them all off at high speed once more. It is worth while to watch a big ant-hill under such conditions. In the sun the whole place swarms with hurrying ants, carrying sticks, caterpillars, or each other, with restless energy. When it gets suddenly cooler they all stop working fast and do everything with painful slowness. The larvæ of a species of locust which periodically undertakes great migrations in the Northern Caucasus has similar reactions, which have been worked out rather carefully by Uvarov.[38] On the first stages of the journey the larvæ march along on the ground in great droves (they do not grow their wings until a later stage in their travels); but they never march at night, and if the temperature falls below about 13° to 15° C. their movements cease and they have to stop wherever they happen to be. In the same way they will stop for only a tiny passing cloud. It is interesting to note that there is also a higher limit of temperature *above* which they will not continue to march, so that they halt sometimes in the middle of the day. We see, then, that this locust is able to remain active only under certain optimum conditions of temperature, and many other examples of the same kind of thing could be given.

3. With many animals the coming of nightfall has precisely the same effect as a cloud over the sun, but the stoppage of work is longer and may require the taking of more elaborate precautions. And in addition we find that another set of animals adapted to a different set of conditions comes out and takes the place of the others. If the changes in conditions are greater, or last longer, many animals migrate away altogether to a more suitable locality or else tide over the bad period in some special way ; for instance, by renouncing all outside feeding and living upon their own fat like a hibernating marmot.

Animal communities are therefore organised into a series of separate smaller communities, each of which is in action at a different time. There are " day and night shifts," wet and dry weather sets of animals, communities of winter and summer, and so on. It would be wrong to get the impression that these time-communities are quite separate from each other. Such is obviously not by any means the case. The point is that the main community changes in personnel to a very large extent at different times ; although the changes are not complete, they are very considerable, and in some cases (*e.g.* day and night) there may be very few species of animals which live in both communities. The community of animals living in one place still remains a definite and fundamental unit, since its periodic changes are regular and characteristic. We might compare the place-community to an elaborate piece of machinery (*e.g.* a motor-car), which still remains one unit although there are several more or less independent special mechanisms contained inside it.

We will now consider some of the time-communities in more detail.

4. *Day and Night.*—In woods, the separation of animals into day and night species may be very easily studied. Take an English oak wood as an example. In the daytime there would be, among a host of other animals, birds like sparrow-hawks, blackbirds, thrushes, woodpeckers, and also bank-voles, weasels, butterflies, bees, and ants. At sunset there is often a short pause when the diurnal animals have gone to rest or begun to think of doing so, and the nocturnal ones have not yet got up full steam. Then there would begin to appear the forerunners of a host of night animals : nightjars, owls, moths, bats, long-tailed field mice, etc. Not only is one kind of animal replaced by another, but one kind of food-chain is replaced by another, and certain niches which are unused by any animal during the day become occupied at night. The weasel—bank-vole industry is changed into a tawny-owl—wood-mouse industry. The woodpecker—ant connection has no equivalent at night, while the moth—nightjar-or-bat chain is almost unrepresented by day. In fact, one food-cycle is switched off and another starts up to take its place. With the dawn the whole thing is switched back again. The two communities of animals are not completely separate, owing to the fact that some animals are not so particular about their

time of feeding ; or else they come out chiefly at dusk and so form a transition from one to the other. Also many animals have regular habits which do not correspond exactly with day and night, owing to the fact that the thing controlling them is not light or heat but something else, such as rain or other weather conditions.

5. Such day and night changes are not found in free-living animals only, but also exist among parasites of mammals, and probably of birds too. Owing to the fact that most mammals sleep either by day or by night there exist corresponding rhythmical changes inside their bodies, especially in temperature. Both in birds and mammals the body is slightly colder during sleep than when they are awake. This rhythm depends entirely upon the activity of the animal, since nocturnal birds like owls have the normal rhythm reversed (*i.e.* they are warmer at night), and this in turn can be reversed by changing the conditions under which they live so as to cause the birds to come out by day and sleep by night. Now there are certain round-worms (nematodes) parasitic in man which show the effects of the sleep rhythm in a very remarkable way. The first species (*Filaria bancrofti*) lives as an adult in the lymphatic glands of man in tropical countries, but its larvæ live in the blood. In the daytime these larvæ retire to the inner parts of the body, mostly to the lungs ; but at night they issue forth into the peripheral circulation, appearing first about five to seven in the evening, reaching a maximum about midnight, and disappearing again by about seven or eight in the morning. This rhythm can be reversed if a person stays up all night and sleeps in the day, which shows that the nematode's activity is affected by rhythmical changes in the conditions of the body like those which we have described above. Another species of Filaria (called *Loa loa*) has larvæ which live in the blood of man, but unlike the other species these larvæ come out only in the day, disappearing at night. It is stated that this periodicity is *not* affected by reversal of sleep, but presumably it must originally have been caused by some rhythm in the bodily environment.[139] A third species has larvæ in the blood which occur in the peripheral circulation equally

by day or by night. The habits of these larval worms have a very important bearing upon the means of transmission from one man to another ; for *F. bancrofti* is transmitted by blood-sucking mosquitos which fly at night, while *Loa loa* is now known to be transmitted by Tabanid flies (*Chrysops dimidiatus* and *silaria*) which bite by day.[59]

6. There are a good many other instances of the day or night habits of blood-sucking insects having an immense influence upon the spread of disease. This has been especially well shown by Carpenter in the course of his studies upon sleeping sickness in the Lake Victoria region in Africa. One big problem was to find out which animal acted as an important reservoir of sleeping sickness from which human beings might become infected. The matter was to some extent simplified by the habits of the tsetse fly (*Glossina palpalis*), which carries the trypanosome of the disease from one host to another. The tsetse is diurnal in habits, and so there are various animals which it never comes across at all in the normal course of events. Certain potential enemies are avoided owing to this ; for it is preyed upon neither by bats which come out at night, nor by tree-frogs, which do not feed except at night.[3b] On the islands of Lake Victoria the most important reservoir animals are the tragelaph (a species of marsh-haunting antelope which comes out to feed just when the flies are " on the bite ") and the hippopotamus, which, although mainly nocturnal, comes out about half an hour before sunset and so is just in time to be bitten by the flies.

7. The most violent fluctuations in light, temperature, and humidity are probably those found in deserts, where a man may be nearly dead with heat in the middle of the day and nearly freezing at night. Often the conditions are so severe that small rodents (which are usually rather sensitive to a dry atmosphere) are able to come out only at night.

A good account of these changes is given by Buxton in his fascinating and scientific book *Animal Life in Deserts*,[2] and the subject has been further studied by Williams[41] in a series of papers on the climate of the Egyptian desert. Williams found, like most biologists who are engaged upon intensive ecological work, that the routine observations taken by meteorologists were not always of much use in the study of animals. Their observations are taken at rather arbitrary times and under extremely unnatural conditions, and are therefore often of little value to the ecologist. To take a simple example, meteorological screens are usually fixed at a height of 4 feet from the ground and the instruments in them record the climate at a height where comparatively few animals live. Furthermore, very few animals live *in the open* at that height—except cows and zebras and children and storks and certain hovering insects. What we have said is particularly true when the communities of animals at night and in the day are being worked out. As a matter of fact, the only kind of data which are of any use in the solving of this kind of ecological problem are accurate charts of temperature, humidity, and rainfall, obtained from continuously recording instruments placed actually in the habitat which is being studied. In most places only the first of these is available, and even that may be absent.

8. A careful study of the changes in external conditions during the day and night with reference to corresponding changes in the activities of animals is very badly wanted, for our ignorance of the matter is profound. It is remarkable to reflect that no one really knows why rabbits come out to feed only at certain times, and on different times on different days. Weather and diurnal changes are no doubt partly responsible, but there our knowledge ends. And yet rabbits are common animals and of great practical importance, and millions of people have watched their habits. We do not know whether light, temperature, humidity, or something else determines the appearance and retirement of animals at certain times. About the food-relationships of nocturnal animals we know less than about those of animals which come out in the day, and that is to say we know pathetically little. And, after all, it is quite as important to have information about the factors which limit animals in time as those which limit them in their spatial distribution, from whatever point of view we regard the question, whether from that of evolution or of wider problems in ecology.

9. Most animals have more or less definite migratory movements during the twenty-four hours of day and night, and in some cases these are regular and rhythmical, but not necessarily correlated exactly with light and darkness. The result of these movements is to alter the composition of animal communities in any one place.   Sanders and Shelford [28] found that among the animals of a pine wood in North America there was a certain amount of diurnal migration up and down in a vertical direction. For instance, one species of spider (*Tetragnatha laboriosa*) was to be found among low herbs at 4.30 a.m. and among shrubs at 8.30 a.m., while another species (*Theridium spirale*) occurred in trees at 4.30 p.m. and in herbs at 8.30 p.m.   Many insects (especially flies) occurred at different heights in the vegetation, depending upon the time of day.   There exist similar vertical migrations among plankton animals in fresh-water lakes and in the sea. A number of species of, *e.g.*, crustacea come nearer to the surface during the night.

10. We may repeat here that the distinction between day and night communities is not necessarily a very sharp one, and that there are a number of animals which come out both by day and by night (*e.g.*, the common black bear of North America [58]), and others whose time-limits are determined by other factors (*e.g.* the slugs mentioned later on). The length of dusk varies throughout the year ; in England it is longest at midsummer and midwinter, and shortest in spring and autumn. Again, the amount of light at night is tremendously influenced by the state of the moon and the occurrence of cloudy weather.   In fact, the distinction between day and night communities may turn out to be less marked than we might at first sight suppose ; but enough has been said to show that the alternation of day and night communities is a very important phenomenon and that it affects animal society profoundly in nearly all parts of the world.

11. In the polar regions there is no such alternation of day and night except during the spring and autumn ; and, since at these times the temperature is too low or the ground too snowy to support much animal life, the species living there are

nearly all typical daylight ones. And these form a permanently working community which lives in continuous daylight during the summer, and may in some cases have very little rest for three months—at any rate, as a population. Conversely, below a certain depth in the sea, or in big lakes, and in subterranean waters, and inside the bodies of animals, there is continuous darkness, so that the animals living there also form homogeneous and permanent communities. Sometimes, however, the bodies of animals reflect the rhythm of their outer environment and cause corresponding differences in their parasite fauna, as in the case of the *Filarias* already described. Probably the most conservative, smooth-working, and perfectly adjusted communities are those living at a depth of several miles in the sea ; for here there can be no rhythms in the outer environment, such as there are on land.

12. As we pass from the poles to the equator the night fauna begins to appear and becomes gradually more elaborate and important, until in such surroundings as are found in a tropical forest it may be more rich and exciting and noisy than the daylight fauna. Alexander von Humboldt [10a] gives a good idea of this. Camping in the Amazon forests in the early nineteenth century, he wrote : " Deep stillness prevailed, only broken at intervals by the blowing of the fresh-water dolphins. . . . After eleven o'clock such a noise began in the contiguous forest, that for the remainder of the night all sleep was impossible. The wild cries of animals rung through the woods. Among the many voices which resounded together, the Indians could only recognise those which, after short pauses, were heard singly. There was the monotonous, plaintive cry of the Aluates (howling monkeys), the whining, flute-like notes of the small sapajous, the grunting murmur of the striped nocturnal ape (*Nyctipithecus trivirgatus*, which I was the first to describe), the fitful roar of the great tiger, the Cuguar or maneless American lion, the peccary, the sloth, and a host of parrots, parraquas (*Ortalides*), and other pheasant-like birds."

In temperate countries the night-life of animals is by no means so abundant or complex. This may be partly due to the fact that whereas the tropical day and night are always twelve hours long throughout the whole year, the night in, say, the south of England is only eight hours long at mid-summer, and the day, therefore, sixteen hours. A consideration of the time-factor in animal communities opens up a number of interesting lines of inquiry ; we have considered day and night in some detail, as it is a clear-cut phenomenon and a fair amount is known about it in a general way. We may now turn to the subject of tidal variation.

13. *Tides.*—In the intertidal zone on the sea-shore there is a marked division of the animals into those which come out or become active at high tide when covered with water and those which appear at low tide. The former group is of course the bigger, and forms the main part of the population, consisting of typical marine species. We can only give a few examples of individual cases, owing to lack of fully worked-out data on the subject. Since most of the dominant animals depend for their living upon plankton organisms in the water, they simply close down at low tide and start feeding again when covered by water at high tide. But there are a number of important animals commonly found on the shore at low tide—mostly birds such as waders, and these differ as to actual species according to the type of shore habitat. It will often be noticed that shore birds divide themselves up into rough zones when they are feeding, some feeding at the edge of the water, others nearer the shore, while others haunt the upper part of the shore near the drift-line. Again, there are different species found on mud-flats, sandy shores, and rocky coasts. These differences in habitat of the birds are no doubt correlated with differences in the food, etc., in the various shore habitats. Besides birds, there are a number of insects which live between tide-marks, *e.g. Anurida maritima*, which during high water hides in crevices in the rocks, surrounded by a bubble of air, and comes out to feed at low tide.[141] The number of insects, mites, etc., which behave like this increases as we go towards the higher parts of the shore, since in these places they do not have to withstand such a long immersion in the sea during high tides.

Sometimes the effect of the tidal rhythm is overridden by that of light and darkness. It appears that many of the corals which form reefs in the tropics only become active and feed at night, closing down during the day.[107b] The times at which they can feed at night will sometimes be conditioned by the state of the tides.

14. *The Weather.*—Most people are aware nowadays that variations in the weather from day to day are caused by the rather irresponsible movements of centres of high pressure (anticyclones) and centres of low pressure (depressions or storm-centres) in the atmosphere. It is customary to speak as if the controlling factors in weather were the depressions; but this is only a convention which owes its origin to the natural belief that anything which disturbs our peace and happiness by bringing bad weather is an actively interfering agent, probably the Devil himself. In reality, the two kinds of pressure areas and their complex relations are equally important, but for convenience we talk in terms of depressions. Changes in weather are associated with depressions travelling over the country, and although the actual path of depressions cannot yet be predicted with certainty, there is a perfectly definite and predictable series of events which accompanies their passage. Generally speaking (in England), a depression produces a zone of rainy weather in its front, and a still larger zone of cloudy weather including and extending beyond the rain-area, while in its rear there is fine weather again. Owing to various complicated factors this ideal series of events is by no means always realised in practice, but the sequence is true on the whole. These changes are accompanied by corresponding changes in temperature and humidity of the air, and by variations in the muddiness, hydrogen-ion concentration, etc., of fresh water.

15. These cycles of weather are of varying length, but are usually of the order of a few days or a week or two, so that they fall in periodicity between tidal or diurnal changes, and the annual cycle of the seasons. They have important effects on animals. Many species are restricted in their activity to certain types of weather. For instance, most mammals avoid rain

because it wets their fur, and, by destroying the layer of warm air round their bodies, upsets their temperature regulation and makes them liable to catch cold. Mice tend to stay at home when it is raining hard, and the badger has to lie in the sun to dry himself if he happens to get wet. Birds are not quite so much affected by weather conditions, since the architecture and arrangement of their feathers usually act as a more efficient run-off for rain. Many insects also necessarily stop work during wet weather owing to the danger of getting their wings wet, or to the drop in temperature often associated with the rain.

16. There are, on the other hand, certain animals which come out only when it is wet (either when is is raining or when the ground is damp after rain). Slugs form a good example of this class of animal. This rule does not apply to all slugs, for there are some species which always live in damp places, as under vegetation ; it holds good mainly for certain of the larger wide-ranging slugs. A record of the activity of slugs was kept for some weeks in woods near Oxford, when the writer was trapping mice. The slugs visited the traps for the bait, and every morning the number of slugs was counted. The number of slugs walking abroad fluctuated greatly, and appeared to be determined mainly by the rain or dampness of the ground. These results were confirmed by casual observations on the general occurrence of slugs on different days. On some days slugs might be seen practically waiting in queues trying to get into the mouse-traps, while on other days the latter would be entirely deserted.

17. These examples serve to illustrate the general idea that animal communities vary according to the weather conditions, and that the variations follow a comparatively regular sequence, although the actual times and periods of the cycle are irregular. We have, broadly speaking, communities of fine weather, of wet weather, and of drying-up weather, but they grade into one another to a large extent. One significant thing is that the weather-cycle may entirely override that of day and night, as in the case of certain slugs. One big black slug with a grey stripe down its back (*Limax cinereo-niger*) comes out in wet

conditions only, and is more or less unaffected by the light or darkness.

18. The effect of weather upon the habits of animals has a certain practical importance, since weather conditions control the habits of many blood-sucking flies, and their disposition to bite people. The tsetse fly (*Glossina palpalis*), which conveys sleeping sickness to man by its bites, is entirely a diurnal feeder, but also shows a marked tendency to bite more on some days than on others. Carpenter [4b] says : " The time when they are most eager to feed is early in a morning after a little rain, when the sun is hardly through the clouds, and it is close and still," while, on the other hand, " if one wishes to avoid being bitten [the time] is in the middle of a day on which there is a fresh breeze, cloudless sky, and brilliant sun."

19. *Seasons of the Year.*—It is unnecessary to say very much about this subject since, although its importance in ecology is immense, the general facts of annual changes and their effects upon the fauna are so well known as to require no underlining. In temperate regions the annual changes in plants and animals are primarily caused by variations in the amount of heat and light received from the sun. In sub-tropical countries this is also true, but there are often in addition very big variations in the rainfall, which are much more abrupt and regular than in more temperate climates. In the actual equatorial belt, with rain-forest, the temperature may be practically the same all the year round, while rain may be the only important climatic factor in which there are marked annual changes. In the Arctic regions the winter is so cold that active life among land animals (other than warm-blooded ones) is confined to the short summer season. It seems probable, then, that the greatest *richness in variety* of com-munities found at different times of year is in the temperate and subtropical regions, although the actual *profusion of species* is not so great as in the tropics.

20. The difference between winter and summer in a country like England is sufficiently great to change the animal com-munities to a considerable extent. This is a matter of every-

day experience, but the details of the changes, and the reasons for them, are at present little understood. A study of seasonal changes in the fauna leads us on directly to a number of problems, one of the biggest of which is that of bird migration—a vast subject which presents us with a number of smaller problems which are essentially ecological. The arrival of certain groups of birds in the spring, e.g. warblers, and their departure again in autumn, are linked up with the summer outburst of insect life, which in turn depends directly upon the rise in temperature, and indirectly upon temperature and light changes acting through plants. For instance, one of the biggest key-industries in many animal communities on land is that formed by aphids which suck the juices of plants. Many small birds depend for their food either directly upon aphids, or indirectly upon them through other animals. The aphids which form this food-supply are only abundant during the height of summer (June, July, and August), and thus their seasonal occurrence has enormously important effects upon the birds. So far the attention of ornithologists has been directed to the accumulation of facts about the actual dates and routes of migrations. This work has resulted in the setting of a number of problems, and the asking of a number of questions. A further advance which will throw light upon the ultimate reasons for the migration-behaviour of birds must be sought along ecological lines, and will only be attained by a careful study of the relations of each species to the other animals and plants amongst which they move in nature, and upon which they are vitally dependent.

21. Since the biological environment is constantly shifting with the passage of the seasons, it follows that the food habits of animals often change accordingly. In the case of many higher animals, a different food is required for the young. The adult red grouse feeds upon the shoots of heather, but the young eat almost entirely insects and other small animals. The food of adult animals also changes in a regular way, especially when they are omnivorous or carnivorous. Niedieck [72] describes how the big brown bear of Kamskatka varies his diet as the seasons pass. When he comes out in

spring from the snow-hole in which he has been hibernating, he has at first to eat seaweed, and a little later on may be seen actually *grazing*. In the middle of June the salmon start to come up the rivers from the sea, and from this time onward salmon forms the bear's staple diet. In August he also eats large quantities of wild peas which are then abundant, and in September, berries. Finally, in the late autumn, he goes and digs up ground-marmots (susliks) in the hills. Having accumulated enough fat to last him through the winter he retires into hibernation again and lives on it in a comatose condition until the following spring.

Many foods are available only at certain times of the year, and this results in the formation of certain temporary niches which may be at the same time world-wide in their distribution. For instance, in Britain the raven (*Corvus corax*) feeds in spring upon the placentæ, or afterbirths, of sheep,[140] and on the Antarctic ice-pack MacCormick's Skua (*Megalestris MacCormicki*) eats the afterbirths of the Weddell seals.[32a]

22. On the other hand, the difference of physical and chemical conditions may be so great that the same niche is filled by different species, often of the same genus, at different times of the year. There are a number of copepods of the genus *Cyclops* which live commonly in ponds, feeding upon diatoms and other algæ. In winter we find one species, *Cyclops strenuus*, which disappears in summer and is replaced by two other species, *C. fuscus* and *C. albidus*, the latter, however, disappearing in the winter. There are also, however, some species of *Cyclops* which occur all the year round, *e.g. C. serratulatus* and *viridis*. Instances of this kind could be multiplied indefinitely, not only from crustacea, but also from most other groups of animals, and many will occur to any one who is interested in birds.

23. We started with the conception of an animal community organised into a complicated series of food-chains of animals, all dependent in the long run upon plants, and we showed that each habitat has its characteristic set of species, but at the same time retains the same ground plan of social organisation. When the factor of time is introduced it is immediately

seen that each place has several fairly distinct communities (distinct in characteristic composition of species, not in the sense of possessing entirely different species) which come out and transact their business of feeding and breeding at different times. We have further seen that the changes in the environment which cause this division into communities at different times are in many cases regular and rhythmical, so that it is possible to classify the latter into definite types—day and night, high and low tide, wet and dry weather, winter and summer, and so on. In spite of the comparatively regular nature of these changing communities, they make the study of this side of ecology excessively complicated, and it is almost impossible to work out the food-cycles, etc., of any ordinary well-developed community of animals with anything remotely approaching completeness. The field worker is faced with masses of preliminary routine work in the way of collecting, etc., with little chance of getting on with a more fundamental study of the problems he is continually coming across. The incredibly intricate and complex nature of a fully developed community of animals is a really serious obstacle which has to be faced, especially as most ecologists are unlikely to be able to obtain the help of more than one or two others.

24. It is therefore desirable, and in fact essential, that any one who intends to make discoveries about the principles governing the arrangement and mode of working of animal communities, should look round first of all with great care, with the idea of finding some very simple association of animals in which the complications of species and time changes are reduced to a minimum. The arctic tundra forms just such a habitat. Here time-changes are practically ruled out (*i.e.* there is only one community in each habitat, and it forms a comparatively homogeneous unit). Experience has shown that it is quite possible for one person to study with reasonable ease the community-relations of arctic animals, in a way that would be entirely impossible in some more complicated place like a birch wood.[25] In our own latitudes simple communities may also be found in certain rather peculiar habitats such as brackish water and temporary pools ; and it seems certain that

our knowledge of the social arrangements of animals will be most successfully and quickly advanced by elaborate studies of simple communities rather than superficial studies of complicated ones.

In order to avoid misunderstanding it should be pointed out that the latter type of work is of great value and interest in other ways, above all for the light that it throws upon other aspects of ecology, e.g. distribution of species, or ecological succession. The statements made above apply only to ecologists who intend to study the principles of social relations among animals, and it must not be thought that the value of general biological surveys is in any way depreciated. The latter will always form a most important part of ecological work in the field.

25. If we followed the subject of time-communities to its logical conclusion (which happily we shall not do, since it would involve a consideration of astronomy and the causes of ice ages, and finally a discussion of the evolution of man), we should have to consider the larger periodic variations in climate from year to year, which undoubtedly exist, even though opinions may differ as to their exact cause and periodicity. For instance, in England there were severe droughts in 1899, 1911, and 1921. In the same way there have been extremely wet years, or very cold winters (see p. 131), all of which have enormous effects upon wild animals. These periodic variations in the climate and weather have chiefly an influence upon the numbers of animals by encouraging or discouraging their increase, and therefore to some extent their distribution.

26. The exact limits of the ranges of a number of animals are constantly shifting backwards and forwards, ebbing and flowing as the outer conditions change, and as the numbers of each species increase or decrease. We understand at present little about the precise causes of these fluctuations in range; but although the immediate influence at work may often be biotic, many of these changes are no doubt ultimately referable to short-period climatic pulsations, whether irregular or regular. For instance, in certain years there are great influxes into the British Isles of various animals not normally found there, or

only rarely. Well-known examples are the crossbill, the arctic skua, the sand-grouse, the clouded yellow butterfly, and various marine fish. These variations in the numbers of animals, or the arrival of entirely strange species, have a definite effect upon the species-composition of animal communities, but they are less important than such smaller rhythms as the seasons of the year. A consideration of this question leads us on naturally to consider the numbers of animals, and the means by which these numbers are regulated. This is a very big subject. It is also a very interesting one, and less is known about it than about almost any other biological subject. The study of animal numbers will form in future at least half the subject of ecology, and even in the present state of our knowledge it seems worth while to devote two chapters to it. This we shall accordingly do.

CHAPTER VIII

THE NUMBERS OF ANIMALS

The subject of this chapter is an extremely important one, (1) but at present only quite a small amount is known about it. (2) Most people do not realise how immense are the numbers of wild animals now, and even more in the past; but (3) in some parts of the world there are still huge numbers of the larger animals, which enable us to adjust our ideas on the subject. Examples among birds are : (4) aquatic birds on the White Nile, (5) guillemots in the Arctic, and (6) penguins in the Antarctic regions ; while examples among mammals are : (7) zebras in Africa, and, comparatively recently, bison in America, passenger pigeons in America, whales and walruses in the Arctic, and tortoises on the Galapagos Islands, and (8) the animals recently protected in various countries. (9) These examples, from amongst birds and mammals, enable us to form some idea of the colossal numbers of smaller animals (*e.g.* springtails) everywhere. (10) Another way of realising the large numbers capable of being reached by animals is to take cases of sudden and almost unchecked increase, *e.g.* among mice, (11) springboks, (12) insects, water-fleas, or protozoa, or (13, 14) epidemic parasites. (15) Such " plagues " are not uncommon under natural conditions, but are (16, 17) especially striking when animals are introduced by man into new countries. (18) This enormous power of multiplication is not usually given full rein, owing to the fact (19, 20) that every animal tends to have a certain optimum density of numbers, which (21) depends among other things upon the special adaptations of each species, and which (22, 23) applies both to herbivores and carnivores. (24) The existence of an optimum density of numbers leads us to inquire how the numbers of animals are regulated. (25) The food-cycle structure of animal communities forms one of the most important regulating mechanisms, (26, 27, 28) enemies being more important than food-supply as a direct limiting check on numbers of most animals ; while reproductive limitation is also of great importance but is more or less fixed by heredity, whereas the other checks are very variable. (29) We lack, at present, the necessary precise data for working out the dynamics of a whole community, and (30, 31) it is therefore not a simple matter to predict the effects of variation in numbers of one species upon those of other ones in the same community. (32) The existence of alternative foods for carnivores has an important influence in maintaining balanced numbers in a community. (33) Animals at the end of food-chains (*i.e.* with no carnivorous enemies) employ special means of regulating their numbers, *e.g.* (34) by having territory, (35) a system found widely among birds, and (36) probably equally widely among mammals. (37) Some carnivorous animals have means of getting food without destroying their prey, and may therefore encourage increase instead of limiting the numbers of the latter.

1. IN the two chapters which follow we shall point out some of the more important things about the numbers of animals

and the ways in which they are regulated, and show how a great many of the phenomena connected with numbers owe their origin to the way in which animal communities are arranged and organised, and to various processes going on in the environment of the animals. We shall first of all try to give a picture of the enormous numbers of animals, both of individuals and species, that there are in every habitat; then we shall describe the great powers of increase which they possess. This leads on to the question of what is the desirable density of numbers for different animals; and it will be seen that the whole question of the optimum number for a species is affected by the unstable nature of the environment, which is always changing, and furthermore by the fact that practically no animals remain constant in numbers for any length of time. We have further to inquire into the effects of variations in numbers of animals, and into the means by which numbers are regulated in animal communities. The final conclusion to which we shall come is that the study of animal numbers is as yet in an extremely early stage, but that it is one of profound importance both theoretically and practically, and one which can best be studied through the medium of biological surveys and study of animal communities, from the point of view of food-cycles and time-communities.

2. It is rather difficult to realise what enormous numbers of animals there are everywhere, not only in species but in number of individuals of each species. The majority of animals, especially in this part of the world, are small and inconspicuous—it is estimated that at least half the species in the animal kingdom are insects—and their presence is therefore not very obvious. If you ask the ordinary person, or even the average naturalist, how many animals he thinks there are in a wood or a pond or a hedge, his estimate is always surprisingly low. Two boys of rather good powers of observation who were sent into a wood in summer to discover as many animals as they could, returned after half an hour and reported that they had seen two birds, several spiders, and some flies—that was all. When asked how many species of all kinds of animals they thought there might be in the wood, one

replied after a little hesitation " a hundred," while the other said " twenty." Actually there were probably over ten thousand. An exactly similar result is obtained with classes of zoological students who are taken out on field work; and it is simply due to the fact that it requires a good deal of practice to find animals, most of which are hard to see or live in hidden places. It is well known that specialising in one particular group of animals enables a man to spot animals which any one else would miss. It is therefore necessary to allow an enormous margin on this account, when one is trying to get an idea of the actual numbers of animals in the countryside.

3. Living as we do in a world which has been largely denuded of all the large and interesting wild animals, we are usually denied the chance of seeing very big animals in very big numbers. If we think of zebras at all, we think of them as " the zebra " (in a zoo) and not as twenty thousand zebras moving along in a vast herd over the savannahs of Africa. To correct this picture of the numbers of big animals, which were so much greater everywhere in the past, to accustom the mind to dealing with large numbers, and to help in forming some conception of the colossal abundance which is reached everywhere still by the smaller animals, we shall describe some of the places where animals of a convincing and satisfactory size may still be seen in enormous and even staggering numbers. These places are mostly rather out of the way and owe the persistence of their rich fauna either to the existence of natural barriers such as an unhealthy climate or an unproductive soil, or to the sensible game-preserving methods of the natives.

4. In the lower reaches of the White Nile, between the vast swamp of the Sudd and Khartoum, there exist countless numbers of water-fowl, many of them large birds as big as a man. For days it is possible to sail past multitudes of storks, herons, spoonbills, cranes, pelicans, darters, cormorants, ibises, gulls, terns, ducks, geese, and all manner of wading birds like godwits and curlews. Abel Chapman [85a] describes the scene as follows : " The lower White Nile, as just stated, is immensely broad and its stream intercepted by low islands

and sand-banks divided one from another by shallows, oozes, and backwaters. At intervals these natural sanctuaries are so completely carpeted with water-fowl as to present an appearance of being, as it were, *tessellated* with living creatures, and that over a space of perhaps half a mile and sometimes more. These feathered armies are composed not only of ducks and geese but also of tall cranes, herons, and storks, marshalled rank beyond rank in semblance of squadrons of cavalry." This amazingly rich bird-life has been photographed by Bengt Berg,[113] and the reader may be referred to his book, the illustrations of which will show that the description given above is not in the least exaggerated.

5. It would be a mistake to imagine that such an abundance of birds is only to be met with in tropical regions. In the Arctic, where continuous daylight throughout the summer encourages a rich harvest of diatoms and other phytoplankton near the surface of the sea, upon which is based an equally rich community of plankton animals, there are still to be found in some places stupendous numbers of sea-fowl. Guillemots are particularly abundant, for they breed in great colonies on the sides of steep sea-cliffs, where their black and white costume makes them very conspicuous. The writer is thinking of one cliff in particular on which the birds could be seen sitting packed close together on every ledge of rock up to a height of a thousand feet or more. When a gun was fired a few odd hundred thousand or million birds would fly off in alarm, without, however, noticeably affecting the numbers still to be seen on the cliff. The photograph in Plate VII (*a*) will give some idea of the number of birds flying off the cliff, and therefore a remote idea of the number *not* flying off. In the photograph each streak of black represents not one bird, but a small flock of ten to thirty. The noise made by these multitudinous egg-layers resembled that produced by an expectant audience in a vast opera house, twittering and rustling its programmes.

6. In the southern hemisphere the penguin rookeries afford an example similar to that of the guillemots in the north; but the penguins spread out their colonies over the ground

and not up the side of cliffs. If one examines the photographs of Adelie penguin rookeries given in antarctic books of travel (e.g. by Mawson [93] for Macquarie Island), one can get a vivid idea of the numbers of birds involved. Imagine several million short gentlemen in dress clothes (tails) standing about in a dense crowd covering several square miles of otherwise barren country (see photo in Plate VII (b)). Viewed from a height they look like gravel spread uniformly over the land, with dark patches at intervals to mark the areas of tussock grass, which stand out as islands in the general ocean of penguins.

In both these latter examples—the penguins and the guillemots—the birds represent the numbers from a very large feeding area concentrated in one place for breeding purposes; and to this extent they do not give a fair idea of the normal density of the population.

7. If we turn again to Africa, not this time to the rivers but to the open grass plains and savannahs, we shall find in some places vast herds of hoofed animals—zebra, buffaloes, and many kinds of antelopes. These are sometimes amazingly abundant. Percival [12e] records seeing a herd of zebras in close formation which extended for over two miles, and other observers have recorded similar large numbers in the case of other animals. Alexander Henry,[8c] describing the abundance of American buffalo in one place in 1801, wrote in his journal : "The ground was covered at every point of the compass, as far as the eye could reach, and every animal was in motion." A hundred years later the bison was reduced to a small herd kept in a national park.[8a] The fate of the American bison is only one example of the way in which advancing civilisation has reduced or exterminated animals formerly so characteristic and abundant. The bison has practically vanished; the passenger pigeon has completely vanished ; but in 1869 a single town in Michigan marketed 15,840,000 birds in two years, while another town sold 11,880,000 in forty days.[8a] The Arctic seas swarmed with whales in the sixteenth century, but with the penetration of these regions by Dutch and English whalers the doom of the whales was sealed, and in a hundred and

fifty years they had nearly all disappeared, while a similar fate is now threatening those of the southern hemisphere. In the photograph in Plate VIII can be seen the skulls and bones of a huge colony of walruses on Moffen Island, up in latitude 80° N., which were slaughtered as they lay there, by these same early explorers. On the Galapagos Islands there is a similar cemetery of giant tortoises, of which only the shells are left to mark their former abundance.[36d] Almost everywhere the same tale is told—former vast numbers, now no longer existing owing to the greed of individual pirates or to the more excusable clash with the advance of agricultural settlement.

8. It is not much use mourning the loss of these animals, since it was inevitable that many of them would not survive the close settlement of their countries. The American bison could not perform its customary and necessary migrations when railways were built across the continent and when the land was turned into a grain-producing area.[8a] Our object is rather to point out that the present numbers of the larger wild animals are mostly much smaller than they used to be, and that the conditions under which the present fauna has evolved are in that respect rather different from what one might imagine from seeing the world in its present state. At the same time there is in many cases no reason why animals should be reduced in numbers or destroyed to the extent that they have been and still are. From the purely commercial point of view it often means that the capital of animal numbers is destroyed to make the fortune of a few men, and that all possible benefits for any one coming later are lost. Enlightened governments are now becoming alive to this fact, and measures are being taken to protect important or valuable animals. Thus the fur seal on the Alaskan Islands, which was in some danger of being gradually exterminated, has increased greatly under protection. Since 1910 killing has been prohibited on the Pribiloff Islands, except by Federal agents, and the herd of seals had increased from 215,000 in 1910 to 524,000 in 1919.[21] Again, the Siberian reindeer has been introduced into Alaska, with similar favourable results in repopulating the country

with animals. A little over a thousand animals were brought over in the period from 1891 to the present day, and the multiplication of these, under semi-wild conditions, has resulted in a great increase. It was estimated that there were 200,000 of them in 1922.[46] It seems, then, that man is beginning to rectify some of his earlier errors in destroying large and interesting animals, and that the future will in certain regions show some approach to the original condition of things before man began to become over-civilised.

9. When we turn to the smaller animals such as insects, worms, etc., we find that there are not very many accurate data about the density of their numbers, but it may be safely said that the numbers of most species are immensely great, reaching figures which convey little meaning to most people. Censuses which have been taken of the soil fauna at the Rothamsted Experimental Station give some idea of the density of numbers reached.[44a] In an acre of arable land there were estimated to be over 800,000 earthworms (these figures being obtained by taking a series of small samples, making complete counts, and then estimating the total number in an acre). In a similar plot of arable land there were nearly three million hymenoptera, one and a half million flies, and two and a third million springtails. These census figures bring out the interesting fact that many groups of small animals which are usually ignored or unfamiliar to zoologists bulk larger in numbers than other groups which have received a very great deal of attention. Such groups are the springtails, of which there were about two and a third millions, while of lepidoptera there were only thirty thousand individuals.

10. We have attempted to give some idea of the great numbers of many large animals in the past; they are still to be found at the present day in secluded parts of the earth. By thinking in terms of large, interesting, and even spectacular species, it is possible to accustom the mind to dealing with the vast numbers in which the smaller, less noticeable, but none the less important forms are nearly everywhere found. We may also look at the matter from another point of view. We can consider what would happen if any one species were allowed

to multiply unchecked for several years. Although various interesting calculations have been made about what would happen if unlimited increase took place, and alarming pictures have been drawn of an earth entirely peopled with elephants so closely packed together that they would be unable to sit down except on each others' knees, there are as a matter of actual fact a number of wild animals which habitually multiply for short periods almost at the maximum rate which is theoretically possible. It is therefore unnecessary to give rein to the imagination in this matter, since we can obtain actual examples. Some small mammals increase in numbers for several years at a very high speed until they reach such an immense abundance that malignant epidemic diseases break out and wipe out the major part of the population, and those which are left start again on another cycle of increase. Mice do this. Such an over-increase of mice has been described very vividly by Holinshed,[48] who wrote of a mouse plague in 1581: "About Hallontide last past, in the marshes of Danesy Hundred, in a place called South Minster, in the county of Essex . . . there sodainlie appeared an infinite number of mice, which overwhelming the whole earth in the said marshes, did sheare and gnaw the grass by the rootes, spoyling and tainting the same with their venimous teeth, in such sort that the cattell which grazed thereon were smitten with a murraine and died thereof." In 1907 another such "plague" occurred in Nevada, during which 15,000 out of 20,000 acres of alfalfa were completely destroyed.[49] The natural increase of the field-mice (*Microtus*) was so terrific that the ground was in many places riddled with holes for miles. A Frenchman, describing a similar outburst in Europe, said that the ground was so perforated with holes as to resemble a sieve. In Nevada it was estimated that there were some 3,000 birds of prey and carnivorous mammals at work in the "plague" district, that these would be destroying about a million mice or more every month, and that this made no appreciable difference to the numbers. Again, one and a half million mice were killed in a fortnight in one district in Alsace during a great outbreak in 1822,[115] while during the mouse plague in 1917 in

Australia 70,000 mice were killed in one stackyard in an afternoon ! [47]

11. In former years great hordes of a small antelope called the springbok or trekbok used to occur periodically in South Africa. [50] These appeared at intervals from the region of the Kalahari Desert, and in some of their migrations they marched south into the settled districts, doing great damage to crops on their travels. Eye-witnesses of these migrations have described the fantastically large numbers of animals taking part in them. One observer, after careful estimation, thought that there were half a million animals in sight at one moment, and it could be shown that the area covered by the whole migrating horde occupied a space of country one hundred and thirty-eight by fifteen miles. Even though they were not equally dense throughout, there must have been a good many ! Another says : "One might as well endeavour to describe the mass of a mile-long sand dune by expressing the sum of its grains in cyphers, as to attempt to give the numbers of antelopes forming the living wave that surged across the desert in 1892 and broke like foam against the western granite range. I have stood on an eminence some twenty feet high, far out on the plains, and seen the absolutely level surface, as wide as the eye could reach, covered with resting springbucks, whilst from over the eastern horizon the rising columns of dust told of fresh hosts advancing." [50]

12. The results of unchecked increase are also seen in a striking way in the big migrations of locusts and butterflies which have been recorded in various parts of the world. (In some of these cases the unusually large numbers have probably been due to local concentration into migratory swarms, rather than to over-increase of the population by breeding. In the majority of cases, however, there is almost certainly over-population, caused either by excessive increase of animals or by unusual scarcity of food, etc.)

13. There are numerous cases in which similar outbursts in numbers among still smaller animals have been sufficiently large to attract notice. In Switzerland the railway trains are said to have been held up on one occasion by swarms of

collembola or springtails, which lay so thickly on the lines as to cause the wheels of the engines to slip round ineffectually on the rails. The fact that individual springtails are usually about one-twentieth of an inch long will give some idea of the numbers involved. Again, a huge multiplication of water-fleas (*Cladocera*) took place in the Antwerp reservoirs in 1896; the numbers were so serious that six men had to work night and day removing the water-fleas by straining the water through wire gauzes. It was estimated that ten tons of water-fleas were taken out—that is, two and a half times the weight of a large hippopotamus.[51] In the sea there are sometimes " plagues " of protozoa. Peridinians (*e.g. Gonyaulax*) sometimes turn the sea to the colour of blood with their vast numbers, off the coast of India, of California, and of Australia.[52] They may be so numerous as to remove most of the free oxygen from the water, so that the fish die from suffocation. Gran once found that the water in Christiania Fjord was milky with a species of coccosphere (*Pontosphaera Huxleyi*), which is a microscopic plant; and estimated that there were five to six million per litre.[52]

14. Finally, there are the diseases caused by various parasitic animals ; these are nothing more than a breaking away of parasites from the control of the host and increasing at an enormous speed. For example, malaria in the blood, and sleeping-sickness, and all such diseases are the result of over-increase of parasites, just as mouse-plagues are the result of over-increase in mice. The most striking epidemic diseases are of course caused by bacteria or by invisible " viruses," but they illustrate the same idea.

15. We started to describe these examples of enormous multiplication in wild animals in order to emphasise the tremendous powers of increase possessed by them and by all animals. Any species, if given the opportunity, is capable of increasing in the same alarming way as the mice, the locusts, or the *Gonyaulax* ; and as a matter of fact most species probably do so occasionally, producing plagues which are rather sudden in onset, and which are terminated by disease or some other factors, or else are relieved by migration during which the

animals mostly perish. It is not a rare or exceptional thing for a species to break out of control of its normal checks ; and we shall have to return to this subject again later on, since we shall see that " plagues " of animals are an inevitable consequence of the way in which animal communities are arranged and of the great instability of the environment.

16. Many of the most striking cases of sudden increase in animals occur when a species is introduced into a country strange to it, in which it does not at first fit harmoniously, often with disastrous results to itself or to mankind.

The most familiar example of such an introduction is that of the rabbit in Australia. It was also introduced into New Zealand, where it multiplied so excessively as to eat down and destroy the grass over wide areas, so that many thousands of sheep died from starvation.[17c] In the same way the Gipsy Moth (*Lymantria dispar*) was introduced into America from Europe, where it became for some years one of the more serious pests in forests, owing to its great increase and spread. It has been shown recently that the increase was due to the absence of its normal parasites, which keep down the numbers in Europe. The introduction of these parasites into America appears to have acted as an effective check, reducing the numbers to reasonable proportions.[94]

17. When an animal spreads rapidly in this way upon being introduced into a new country, there is usually a definite sequence of events, which is rather characteristic. At first the animal is unnoticed for several years, or else is highly prized as forming a link with the home country. Thus the starling was introduced into New Zealand by acclimatisation societies bent upon brightening the country with British birds.[17d] The next stage is that the animal may suddenly appear in the dimensions of a plague, often accompanied by a migration, as when huge armies of rats marched over New Zealand in the early days.[17b] The starling was instrumental in spreading the seeds of the common English blackberry in New Zealand and has been undoubtedly one of the biggest factors in the production of the blackberry plague there. This has resulted in the formation of thickets of blackberry covering

the country for miles, making agriculture impossible, and in some places forming a danger to lambs, since the latter get caught inextricably on the thorns of the blackberry plants. Finally, after a good many years there is often a natural dying down of the plague. This is in most cases not due to the direct efforts of mankind in killing off the pests, but appears rather to be due to the animal striking a sort of balance with its new surroundings, and acquiring a set of checks which act fairly efficiently. Thus the rabbits in New Zealand now apparently have periodic epidemics, which reduce the population. A parallel case among plants is that of the Canadian water weed (*Elodea* or *Anacharis canadensis*), which was introduced into Europe by an enthusiastic botanist during the nineteenth century, and subsequently spread for some years like wildfire, choking up rivers and lakes. After a certain time it appeared to lose its great multiplying power, and has now settled down to be a normal and innocuous member of the flora. The fame of *Anacharis* is still so great, however, that a certain town council in Britain, faced with a plague of "waterbloom" in one of their lakes (water-bloom being caused by species of blue-green algæ increasing abnormally in the water), hopefully stocked the lake with swans in order to eat down the *Anacharis* which was living there quite harmlessly. There are now a great many swans there.

18. We have seen that animals possess extremely high powers of increase, which sometimes have a chance of being realised with results which are often very remarkable. Such high powers of increase do not merely reside in animals which have very large broods or breed very fast: there is no animal which could not (theoretically) increase enormously, given sufficient time and opportunity. Not more than fifty years at the maximum, or in most cases not more than two or three years, are required to achieve this result. For instance, the opossum (*Trichosurus vulpecula*), which was introduced from Australia into New Zealand, has only one young one every year, yet in some places it has increased alarmingly. Thomson [17a] says that the black-tailed wallaby introduced on to Kawau Island " ate out most of the vegetation, and starved

out most of the other animals, being assisted in this by the hordes of opossums. They came out at night in the fields, grazing like sheep, and in the summer went into the garden, stripping it of fruit and vegetables."

As Hewitt has pointed out, the converse of this is also true, and great abundance is no criterion that a species is in no danger of extinction. Just as an animal can increase very quickly in a few years under good conditions, so on the other hand it may be entirely wiped out in a few years, even though it is enormously abundant. The argument that a species is in no danger because it is very common, is a complete fallacy; but is very often brought forward quite honestly, especially by people who have a financial interest in destroying the animals. One might mention the case of whales in the southern hemisphere.

19. The examples of upsets in the normal balance of numbers which we have described bring us up against the question: what is the desirable density of numbers for any one species ("desirable" being used in the teleological sense of that density which will in the long run give the best chances of survival for the species)? The question of the desirable number on a given area has received a great deal of attention from people studying the ecology of human beings. It is found that there is an optimum density of numbers for any one place and for people with any particular standard of skill.[6] To take a simple case: when a man is running a farm he cannot afford to employ *more* than a certain number of men on it, since after a certain point the income he gets from the farm begins to diminish. It does not pay to put in more than a certain amount of work as long as the standard of skill remains the same. If a new invention or a new idea opens up new lines of production, then it becomes possible to employ more men with advantage, but not more than a certain number. On the other hand, it does not pay the owner to employ *less* than a certain number of men if he is to get the maximum return from his outlay. If there are too few men working, the maximum production is not reached.

20. Let us see whether the idea of optimum numbers

applies to wild animals, and whether the analogy with man can be followed up. If we go into the question carefully, it soon becomes clear that there is an optimum density in numbers for any one species at any one place and time. This optimum number is not always the same and it is not always achieved, but in a broad way there is a tendency for all animals to strike some kind of mean between being too scarce and too abundant. As examples we may take the domestic cat in two of its wilder moments. Some years ago a schooner was run ashore on the coast of Tristan da Cunha, a remote island in the Southern Atlantic, and some of the ship rats were able to get ashore and colonise the island. In a few months they bred and increased excessively until they became quite a plague, even attacking and eating rabbits on the island. The inhabitants accordingly introduced some cats with the praiseworthy idea of extinguishing the rats. But the rats were so very much more numerous that they killed off the cats instead.[60] In this case there were *not enough* cats. They were overpowered by weight of numbers.

The other example is also about a small island, called Berlenga Island, off the coast of Portugal.* This place supports a lighthouse and a lighthouse-keeper, who was in the habit of growing vegetables on the island, but was plagued by rabbits which had been introduced at some time or other. He also had the idea of introducing cats to cope with the situation, which they did so effectively that they ultimately ate up every single rabbit on the island. Having succeeded in their object the cats starved to death, since there were no other edible animals on the island. In this case there were *too many* cats.

21. If we follow up further the analogy with human density of population, it becomes clear that every animal tends to have a certain suitable optimum which is determined mainly by the habits and other characteristics of the species in question. But these are continually changing during the course of evolution, and any such change is liable to cause a corresponding alteration in the optimum density of numbers. For instance,

* This incident was related to me by Mr. W. C. Tate, the well-known authority on Portuguese birds, and is published here with his permission.

if the cats on Tristan da Cunha had possessed poison fangs like a cobra they might have been able to maintain themselves with a small population. On the other hand, if the cats on Berlenga Island had possessed chloroplasts like *Euglena*, they might have been able to exist permanently, without eating out their food-supply. Again, the density possible for a species depends partly upon the size of the animals. Given the same food-supply and other things being equal, a small species can be more abundant than a large one. This has a certain importance in ecology, since there are a great many examples of species in the same genus, and with the same sort of food habits, differing in size to a very marked extent. The common shrew (*Sorex araneus*) and the pygmy shrew (*Sorex minutus*), the small and large cabbage white butterflies (*Pieris rapae* and *brassicae*), the smooth newt (*Molge vulgaris*) and the crested newt (*Molge cristata*), are instances. Of course if the size-differences are too great they often automatically involve different food habits, so that the two animals cannot be compared closely.

22. The principle of optimum density applies equally to any herbivorous animal. Under normal circumstances the numbers of deer are kept down by two big factors—enemies and disease. Recently the deer in a sanctuary in Arizona were left to themselves for some years. Owing to the absence of their usual carnivorous enemies (*e.g.* cougars or wolves) they increased so much that they began to over-eat their food-supply, and there was a serious danger of the whole population of deer starving or becoming so weakened in condition as to be unable to withstand the winter successfully. The numbers were accordingly reduced by shooting, with the result that the remaining herds were able to regain their normal condition.[61] Here it was clear that the absence of their usual enemies was disastrous to the deer, that the former are in fact only hostile in a certain sense, in so far as they are enemies to individual deer; for the deer as a whole depend on them to preserve their optimum numbers and to prevent them from over-eating their food-supply.

23. One more example may be given. Carpenter, when

carrying out a survey of the islands on Lake Victoria in order to discover the distribution and ecology of the tsetse fly, noted that islands below a certain size did not support any flies at all, although the conditions for breeding and feeding (which are well defined and regular) were otherwise apparently quite suitable.[5a] The explanation of this was probably that the fly population is subject to certain irregular checks upon numbers, and that any one population must be sufficiently large to survive these checks. There would not be a big enough margin of numbers on a very small island.

There are suggestions of a similar state of affairs among certain protozoa. It has been found that if there are too few individuals in a culture they do not live so successfully,[91] and this is also said to be true of cells growing in tissue-cultures. Again, it has been found that the minimum density is not the optimum density for a population of the fruit-fly *Drosophila* growing in the laboratory.[90] It is quite probable that there are sometimes physiological or even psychological reasons controlling the desirable density of population, just as it is bad for most people to live alone, or, on the other hand, under too crowded conditions. But we do not know much about this matter among wild animals. It may be of very great importance in their lives and cannot be ignored as a possible factor affecting numbers.

24. Before going on, it will be convenient to sum up what has been said so far about the numbers of animals. Most animals are more numerous than is usually supposed, and it is necessary to accustom the mind to dealing with large actual numbers of individuals. One is the more likely to under-estimate the numbers of animals, owing to the destruction of the large and more conspicuous species which were formerly so much more abundant in many parts of the world, now occupied by industrial or agricultural civilisation. Descriptions of the enormous numbers in which these larger animals still exist in the more secluded parts of the world and of the former numbers of animals which are now rare or extinct, enable us to grasp to some extent the vast abundance of the smaller and more inconspicuous forms.

That is the first point—the vast numbers of animals almost everywhere.

We have further seen that all animals possess an extremely high power of increase, which if unchecked leads to over-increase on a large scale, so that a " plague " of one species or another is produced ; the best-known cases being mice and locusts. All animals are exerting a steady upward pressure in numbers, tending to increase, and they sometimes actually do so for a short time. That is the second point. We next considered the question of the desirable density of numbers for a species, and we saw that each species tends to approach a certain optimum density, neither too low nor too high, which is not the same at different times or in different places. If there are too few individuals the species is in danger of being wiped out by unusually bad disasters, and if the numbers are too great other dangers arise, the most important of which is the over-eating of the food-supply. The latter is always the ultimate check on numbers, but in practice other factors usually come in before that condition is reached.

25. We now have to consider the regulation of numbers, the ways in which this desirable density of numbers is maintained. How do animals regulate their numbers so as to avoid over-increase on the one hand and extinction on the other ? The manner in which animals are organised into communities with food-cycles and food-chains to some extent answers the question. As a result of the existence of progressive food-chains, all species except those at the end of a chain are preyed upon by some other animals. Snails are eaten by thrushes, the thrushes by hawks ; fish are eaten by seals, seals by sea leopards, and sea leopards by killer whales ; and so on through the whole of nature. Most species usually have a number of carnivorous enemies, but in some specialised cases may have only one. The latter condition is, however, extremely rare ; it is the commonest thing in the world to find a species preying exclusively upon another, but unusual for a species to have only one enemy. Every species has also a set of parasites living in or on it, which are often capable of becoming dangerous when they are very numerous. So, in a

general way, the food-cycle mechanism is in itself a fairly good arrangement for regulating the numbers of animals, and it works efficiently as long as the environment remains fairly uniform, or at any rate as long as its periodic pulsations continue fairly steadily and regularly. If the balance of numbers in a community is upset by some sudden and unusual occurrence, then the ordinary relations of carnivores and parasites to their prey are no longer effective in controlling numbers, and various results of a curious nature ensue. With the effects of irregularities in the surroundings of animals we shall deal more fully in the next chapter. We are here concerned chiefly with the ways in which the general system of food-cycles and food-chains in animal communities acts as a method of regulating numbers.

26. It is plain enough that the amount of food available sets an ultimate limit to the increase of any animal; but in practice, starvation seldom acts as a direct check upon numbers, although the possibility of it is always present. Instead we find that other factors, such as enemies of all kinds, usually keep numbers down well below the point which would bring the population in sight of starvation. There appear to be several good reasons for this. First of all, food, whether of an animal or plant nature, is not always available; or, what comes to the same thing in the end, is not always increasing to keep pace with the needs of the animals requiring it; so that the maximum numbers feasible for an animal at any moment are not only determined by the food-supply at that moment, but must be adjusted to the needs of the future. It would be an unworkable system for animals to live all the time up to the extreme limits of their food-supply, since no margin would be left for the times of scarcity which are always liable to occur. This can be well seen in the example of deer in Arizona quoted previously, where increase during the summer imperilled the food-supply for the following winter. It is one of the most obvious ideas to all stock-farmers that the number of cattle which can be kept on a given acreage is determined by the margin of food left over for the winter (when plant growth ceases), as well as by the immediate requirements of the

animals during the summer. Another point is that over-eating of the food-supply usually results in the destruction of the entire population, irrespective of individual merits. There have been instances recorded of the various oak moths (such as *Tortrix viridana*) eating all the leaves of the trees upon which they were living and then simply dying of starvation, just as the cats did on Berlenga Island.[142]

27. It is usual, therefore, to find that gregarious gluttony of the whole population is avoided by having various other checks which act in two ways, first by affecting the chances of reproduction or by limiting the number of young produced, and secondly by eliminating the animals themselves. The first method is often fixed more or less permanently by the hereditary constitution of each species ; but there are a great many cases known in which the weather affects mating or breeding, or in which climate or food-supply vary the number of young produced in a brood, or the number of broods born in a year. For instance, the short-eared owl (*Asio flammeus*) may have twice as many young in a brood and twice as many broods as usual, during a vole plague, when its food is extremely plentiful.[114] But these variations in the reproductive capacity are small compared to the limits which are imposed by the constitution of the animals.

The second result is brought about mainly by means of predatory enemies, carnivores or parasites, or both, not to speak of other checks such as climatic factors. These influences dispose all the time of a certain margin of the population, so that there are left a certain number of comparatively well-fed and, as it were, well-trained animals ; for these checks act selectively and probably have important effects on the quality as well as the quantity of the population. Starvation only comes in in various indirect ways, as by lowering the resistance to attack by carnivorous enemies or parasites, or to the weather, and so increasing the selective power of these agencies.

28. The regulation of numbers of most animals would appear, therefore, to take place along the following lines. Each species has certain hereditary powers of increase, which are more or less fixed in amount for any particular conditions. It

is usually also kept down in numbers by factors which affect breeding, e.g. lack of breeding-sites, etc. It is further controlled by its enemies, and if these fail, by starvation. But the latter condition is seldom reached. There are a few species which seem to regulate their numbers almost entirely by limiting reproduction, although they belong to groups which are normally controlled by carnivores. There is a species of desert mouse (*Dipodomys merriani*) which only has two young per year, that is to say, probably very few more than would be necessary to replace deaths in the population caused by old age or accident.[148] This, however, is unusual except in the case of animals at the end of a food-chain, with which we shall deal later.

It has been necessary to speak in generalities, since so little is known at present about the rules governing the regulation of animal numbers. There are, however, a number of special separate phenomena which we shall pick out, and which will serve to illustrate the importance of animal interrelations, and the study of animal communities along the lines of food-cycles.

29. When we are dealing with a simple food-chain it is clear enough that each animal to some extent controls the numbers of the one below it. The arrangement we have called the pyramid of numbers is a necessary consequence of the relative sizes of the animals in the community. The smaller species increase faster than the large ones, so that they produce a sufficient margin upon which the latter subsist. These in turn increase faster than the larger animals which prey upon them, and which they help to support; and so on, until a stage is reached with no carnivorous enemy at all. Ultimately it may be possible to work out the dynamics of this system in terms of the amount of organic matter produced and consumed and wasted in a given time, but at present we lack the accurate data for such calculations, and must be content with a general survey of the process. The effect of each stage in a food-chain on its successor is easy to understand, but when we try to estimate the effect of, say, the last species in the chain upon the first, or upon some other species several stages away, the matter becomes complicated. If A keeps down B, and B

keeps down C, while A also preys on C, what is the exact effect of A upon C? Two examples will show the sort of way in which this process works.

30. For some years the great bearded seal or storkobbe (*Erignathus barbatus*) has been very intensively hunted and killed by Norwegians who go up every summer into the outer fringes of the ice-pack round about Spitsbergen. They seek the seals for the sake of their skins and blubber. The serious toll taken of their numbers can be gauged by the fact that one small sealing-sloop may bring back five thousand skins and sometimes many more in the course of a single summer. In spite of this steady drain on the numbers of seals the animals are, if anything, more abundant than ever. This appears to be due to the fact that the Norwegian sealers also hunt and kill large numbers of polar bears, whose staple article of diet is the bearded seal, which they stalk and kill as they lie out on the pack-ice. By reducing the numbers of bears the sealers make up for their destruction of seals, since there are so many extra seals which would otherwise have been eaten by bears. The diagram in Fig. 9 sums up the situation which has just

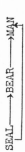

SEAL ⟶ BEAR ⟶ MAN

FIG. 9.

been described. In this case we know the results of man's interference, but they might very well have been different. For instance, if fewer bears had been killed, the seal numbers might have gone down considerably. On the other hand, if the same number of bears had been killed and more seals destroyed by the Norwegians, the seal numbers might also have gone down. The final result, as far as seals are concerned, depends entirely upon the relative numbers and destructive powers of the species concerned. In this case a balance happens to have been struck.

31. The second example illustrates the same point. The tsetse fly (*Glossina palpalis*) is preyed on in part of the Lake Victoria district by a small species of dragonfly (*Cacergates leucosticta*), and the latter is preyed on in turn by a larger

dragonfly. Both dragon-flies are eaten by various species of bee-eaters (*Merops* and *Melittophagus*).[4a] Problem: what is the effect of the bee-eaters upon the numbers of tsetse flies? The diagram in Fig. 10 sums up the food-relations which we

TSETSE ⟶ DRAGONFLY ⟶ DRAGONFLY ⟶ BEE-EATER

FIG. 10.

have described. It is clear that in this case we cannot say at a glance whether the birds are having a beneficial influence by helping to reduce tsetse flies, or the reverse. The result depends entirely upon the relative numbers of the species concerned in the matter, and upon a number of other things, such as the food preferences of the bee-eaters, the number of individuals eaten in a given time by the dragonflies, the rate of increase of the different species, and so on. But the example does show that each species will have some effect upon the numbers of the others, even though we cannot precisely define it without further investigation. In fact, no species in a community, unless it happens to live a very isolated life or be very rare, is without its effect upon numbers of the rest of the community, and that is why it is practically hopeless to reach any complete knowledge of the natural methods of regulation of numbers of an animal without doing a general biological survey, backed up later by some investigation of the food-cycle.

32. It might be thought that there would be some danger of enemies doing *too much* in the way of controlling the numbers of their prey, so that the carnivore would run a risk of eating out its food-supply, and the prey of being exterminated or reduced below its lower limit of safety. There is, however, a natural method by which such a contingency is usually avoided, depending upon the fact that most carnivores do not confine themselves rigidly to one kind of prey; so that when their food of the moment becomes scarcer than a certain amount, the enemy no longer finds it worth while to pursue this particular one and turns its attention to some other species instead. This process was pointed out by Hewitt, who gave

as his example the goshawk (*Accipiter atricapillus*) in Canada, which preys alternately on the varying hare (*Lepus americanus*) and upon grouse, according as one or the other is more abundant. In this way, whenever one species becomes for any reason scarce, the goshawk tends to eat more of the other and so allows the first one a certain amount of respite.[8b] This switch arrangement is common enough in animal communities, and is probably an important factor in preventing the complete extermination of animals which happen for any reason to be at a rather low ebb of numbers (*e.g.* after an epidemic). In just the same way, the red fox in Canada preys on mice or varying hares according to their relative abundance.

33. It will already have occurred to the reader that the animals which are at the end of food-chains—at the top of the pyramids—are in a peculiar position, since they have no further carnivorous animals present which might control their numbers, although they have of course parasites. These animals have in many cases evolved rather curious methods of regulating their numbers, of which we can only mention a few here. The Emperor Penguin is a large bird which forms the end of a long chain of marine animals (it appears to live chiefly on animals like fish and squids), and breeds in the heart of the very cold Antarctic winter. Since it has no serious enemies to control its numbers (nothing is known as to whether it has epidemics), it seems to depend chiefly on climatic factors to bring this about: or rather we should say that the only checks against which it has to produce extra numbers are climatic ones. One important thing is the cold, since the birds attempt to incubate their eggs and hatch their young at a temperature ranging below −70° F. and in severe blizzards. They are also destroyed by avalanches of snow, which cover them and cause desertion and freezing of the eggs. (Some birds were seen attempting to hatch out pieces of ice, which they had mistaken for their eggs.[32b]) In other cases, animals at the end of food-chains may control their numbers by not breeding at all in some years. This appears to happen with the snowy owl, and probably with skuas, in certain years when food (especially lemmings) is scarcer than usual. Or again, the

reproduction of the carnivore may be always adjusted to such a low rate that there is hardly ever any danger of over-eating its food supply, and its numbers always remain relatively small. This is not so common, however, as other methods.

34. The regulation of numbers of terminal animals is seen at its best in some of the birds and mammals which are either at or near the end of food-chains, e.g. hawks and tigers on the one hand, or warblers and insectivorous animals on the other. The fact that animals become less abundant as we pass from key-industry herbivores to the carnivores at the end of the chain makes it possible for animals at a certain point in the series (at a certain height in the number-pyramid) to limit their numbers by dividing up their country (and therefore the available food-supply) into territories each owned by one or a few individuals. For instance, in the English Lake District each buzzard or pair of buzzards requires a certain stretch of country to supply it with enough food, and the same applies to a bird like the Dartford warbler, which lives in the heather and furze heaths of Southern England. The division of country into territories is especially common at the beginning of the breeding season, when it is necessary not only to provide for the immediate needs of the animals, but also for the needs of the young which will appear later on.

35. We owe our knowledge of the existence and nature of bird territory chiefly to Eliot Howard,[9] whose remarkable studies on the subject, especially among the warblers, have opened up an entirely new field of ecological work. He showed that among birds like the willow wren (which migrates northwards into England every spring) there is a very regular system of dividing up their habitat into parcels of land of roughly equal value. The arrangement is that the male birds arrive first in the early spring, before the females, and fight amongst themselves for territory; and are then followed by the females, each one of which becomes attached to one male. Ultimately the nest is built and young produced and reared. At the end of the season the territories are given up and the birds go south again. We are not concerned here with the ways in which this territorial system in warblers and other birds is connected

with courtship and nesting habits. These and other matters of great ecological interest may be found described and discussed in the works of Howard. It should be pointed out that territorial systems among birds are not always for the purpose of dividing up the food-supply in a suitable way. They may also be equally important in limiting numbers by being concerned with nesting-sites.

36. Less is known about territory conditions in carnivorous mammals, but it is pretty clear that they do in many cases have territories, and for the same reasons as with hawks and warblers. It is well known that animals like the African lion or Indian tiger are few in numbers, and that each district will have one pair or one family living in it. We know practically nothing about the way in which such animals settle the size of their territory, and little enough about the extent to which the territorial system is found among mammals at all. There are indications that some herbivorous mammals limit their numbers to some extent in this way. For instance, Collett [33b] says that the lemming does so in normal years. It is probable that insects like ants, which live in great towns, have some system of spacing out their colonies so as to avoid overcrowding. The whole subject requires more investigation before we can say exactly how important a part it plays in the lives of animals ; but Howard's work on birds is sufficient to show that it is certainly a very effective method of limiting numbers in some species, and it will probably be found to occur very widely among animals which do not possess any other convenient means of regulating numbers.

37. We may conclude this chapter by referring to the animals which live by exploiting the work of other species, but do not actually eat them or destroy them in the process. The majority of parasites belong to this class. Such animals as tapeworms do not always harm their hosts very greatly, except that they divert a certain amount of food from its proper destination in the tissues of the host. In the same way fleas, if not too numerous, do not necessarily do much direct harm by withdrawing blood from their host, although they may accidentally spread the germs of disease in that way. Parasites

do not usually limit the numbers of their hosts except when the latter have increased unduly (as when fish die from tape-worm epidemics or grouse on overcrowded moors from nema-tode disease), or when the parasite has got into the wrong host (as with the trypanosome, which causes human sleeping-sick-ness). Most parasites exist by exploiting the work done by their hosts, without actually destroying them, just as a black-mailer takes care not to ask for too much money at one time. Besides parasites there are also certain animals which are not parasites (in the sense of living on their host), but which at the same time employ similar means of obtaining sustenance from their prey without destroying the latter. For instance, some species of ants keep farms of aphids which they visit for the purpose of getting drops of liquid containing food which has not been completely absorbed by the aphids themselves. These aphid farms may be carefully protected by the ants.[143] Another example of avoiding killing the goose that lays the golden eggs is described by Beebe.[36b] On one of the Galapagos Islands there lives a scarlet rock-crab (*Grapsus grapsus*) which inhabits the lava rocks by the seashore. This crab is pursued by a species of blue heron (*Butorides sundevalli*) which catches the crab by the leg, upon which the crab breaks off the leg by autotomy, leaving it in the possession of the heron. Thus both animals get what they want—the heron its food and the crab its life.

These cases of exploitation without destruction have been described in order to show that the food-cycle mechanism is not always effective in limiting numbers of animals. In such cases some other means must be employed. Sometimes another animal in the community, which still follows the old method of destruction, acts as a real control of the numbers of the species concerned.

CHAPTER IX

VARIATIONS IN THE NUMBERS OF ANIMALS

The numbers of animals (1) never remain constant for very long, and usually fluctuate considerably, and often rather regularly, *e.g.* (2) many insects, (3) marine littoral animals, (4) protozoa in the soil, and elephants in the tropical forests. (5) The primary cause of these fluctuations is usually the unstable nature of the animals' environment, as is shown by the effects of periodic bad winters on the numbers of certain small birds. ((6) Fluctuations in numbers are of common occurrence among birds, but their causes are not usually known.) (7) Further instances of the irregular nature of the environment are cyclones in the tropics, and droughts affecting small ponds, but (8) such factors are not necessarily destructive, and may favour sudden increase of the animals. (9, 10) Some of the best data about fluctuations in numbers are from mammals, *e.g.* the lemming, whose fluctuations are partly controlled by climate and partly by periodic epidemics, (11) the latter occurring in a number of other small rodents, such as mice, (12) rabbits and hares, (13) marmots and muskrats; but (14) the beaver forms an interesting exception to this rule. (15) These fluctuations in numbers affect the other animals dependent upon the rodents, and since (16) the length of the period of fluctuation depends chiefly on the sizes of the rodents concerned, the final effects upon animal communities are very complex. (17) Wild ungulates, at least in some cases, also have periodic epidemics. (18, 19) The occurrence of " plagues " of animals results from the structure of animal communities, and from the irregularities of the environment, and (20) especially from the irregularities occurring over wide areas, at the same time. (21) Among other results of these periodic changes in numbers are changed food habits, since (22) food preferences depend both on the quality and the quantity of the food. (23) The other habits of a species may also change with the variations in density of its numbers.

1. So far we have been speaking as if the numbers of animals remained fairly constant. We have been describing the general mechanisms which assist in bringing about the optimum density of numbers for each species. In the present chapter we shall point out that practically no animal population remains the same for any great length of time, and that the numbers of most species are subject to violent fluctuations. The occasional " plagues " of animals already referred to are extreme

examples of sudden variations in numbers. But the variations are usually less spectacular, although hardly less important. If we take any animal about which we possess a reasonable amount of information, we shall find that its numbers vary greatly from year to year. The common wasps (*Vespa vulgaris* and *V. germanica*) are very abundant in some summers, and very scarce in others. It is well known to collecting entomologists that there are very good and very bad years for butterflies. Although the data are scattered and have not so far been properly correlated, a study of the existing literature leaves no doubt that there is enormous annual variation in the numbers of insects. A typical example is recorded by Coward,[62] who noted that the mottled umber moth (*Hybernia defoliaria*) was particularly abundant on oak trees in Cheshire in 1918 and 1919, and especially in 1920, and caused much defoliation. At the same time leaf-galls, especially the spangle (caused by *Neuroterus lenticularis*, a species of the *Cynipidae*), were unusually plentiful in 1920. He also noted that the acorn-crop was average in 1918 and 1919, and failed entirely in 1920. These changes in the trees and the insects must have had an appreciable effect upon the other animals in the oak woods, since jays and wood-pigeons eat the acorns; starlings, chaffinches, tits, and armies of warblers eat the caterpillars of the moth; while pheasants and other birds eat the spangle galls. This example will give some idea of the way in which the numbers of insects are always shifting from year to year, and how the changes must inevitably affect a number of other animals associated with them, often causing the latter in turn to vary in numbers.

2. In other countries similar variations in numbers occur. The sugar-cane froghopper of Trinidad fluctuates comparatively regularly, with a period of four or five years.[152] The bad "cotton-worm" (moth) outbreaks in the United States occur at intervals of about twenty-one years.[116] Aphids occur very numerously in some summers, *e.g.* in 1836 there was a very big maximum of numbers in Cheshire, Derbyshire, and South Lancashire and Yorkshire, and in one place the swarms occurred over an area of twelve by five miles.[81]

There were similar aphid swarms in parts of England in the year 1869,[22] and doubtless many other ones which have escaped being recorded. The aphid increase sometimes affects the numbers of their normal enemies such as ladybirds and syrphid flies, which may attain vast numbers at an aphid maximum.[22]

In some cases the fluctuations in numbers of insects are extraordinarily regular in their rhythm. The most remarkable of these are the cycles in numbers of cockchafers (*Melolontha vulgaris* and *M. hippocastani*) in Central Europe. Every few years these beetles appear in countless numbers, and the heavy damage which they do to crops and forest trees has directed much attention to the phenomenon, so that we possess very good records for a number of years back. In some districts there is a regular three-year cycle, in others a four-year cycle. The maximum numbers have in some places occurred exactly every three years for more than sixty years at a time.[95] There is also a fifteen-year period in the size of the maxima.

3. Marine animals also are subject to considerable variations in numbers from year to year, and these are well shown by the careful notes made by Allee [63] upon the marine littoral species in the neighbourhood of Woods Hole, Massachusetts. An echinoderm, *Arbacia punctulata*, was particularly abundant in 1917, but there were practically none in 1918. By 1919 the numbers had recovered again. The hydroid *Tubularia*, which usually dominates the wharf-pile association in early June, and dies away by early August, was so much influenced by the early season of 1919 that it had entirely gone by early July (instead of August); and in consequence many animals such as the molluscs *Columbella* and *Lacuna*, which feed upon *Tubularia*, were very scarce that year, while the other members of the July *Tubularia* association were also scarce or absent (e.g. polychaetes of various genera). The failure of *Tubularia* to be present in the right month affected the numbers of all the other animals associated with it. 1920 was normal, but a similar upset took place again in 1921.

4. The work which has been done upon soil protozoa at

Rothamsted shows that there are similar fluctuations in numbers among these animals, but upon a much smaller scale. These variations are of the order of two days instead of two years, but the principle is the same.[44b] If we go to the other extreme it can be shown that animals as big as the Indian elephant are also subject to fluctuations in numbers, caused by epidemics ; but at very long intervals, of seventy to a hundred years.[81b, 117] Whenever a group of animals or any one species is studied carefully over a series of years, it is found to vary in numbers in a more or less marked way. There is not space here to give all the evidence for this statement, but it seems to be true in practically all cases where there are any accurate data. Even when a species does not vary very much from year to year, it has in the vast majority of cases a marked variation within each year, caused by its cycle of reproduction. Every year there is an annual increase in numbers from comparatively few individuals, among such animals as protozoa, rotifers and water-fleas, many of which possess some well-defined means of increasing rapidly—by parthenogenesis, for instance. In fact, the numbers of very few animals remain constant for any great length of time, and our ideas of the workings of an animal community must therefore be adjusted to include this fact. It is natural to inquire why the numbers vary so much, and why, with all the delicate regulating mechanisms described in the last chapter, the community does not more successfully retain its balance of numbers.

5. The chief cause of fluctuations in animal numbers is the instability of the environment. The climate of most countries is always varying, in some cases regularly—as in the case of the eleven-year cycles in temperature and the frequency of tropical cyclones associated with the sun-spot cycle ; or of snowfall in Norway, which has a very marked short periodicity of three or four years.[23, 24] On the other hand, there are a number of irregular and so far unpredictable cycles, such as that of rainfall in England. The variations in climate affect animals and plants enormously, and since these latter are in intimate contact with other species, there are produced further disturbances which may radiate outwards to a great

distance in the community. Let us take the example of small birds in England. At intervals of ten or more years (and sometimes less) there have occurred in England very severe winters accompanied by continuous frost and snow for several months, which have killed off a very large proportion of the smaller birds each time. Such winters, resulting in the death on a large scale of small birds like thrushes, blackbirds, and tits, are recorded as having occurred in the years 1111, 1115, 1124, 1335, 1407, 1462, 1609, 1708, 1716,[81a] 1879,[118] 1917. Undoubtedly there are many more which have escaped record. In 1407 there was a very long and severe winter, with frost and snow during December, January, February, and March. Thrushes and blackbirds and many thousands of smaller birds died from hunger and cold. In 1716 numbers of goldfinches and other species were destroyed.

Recently there was a very severe winter in England (1916–17) which caused much death among birds and lowered the numbers so seriously that they have only just recovered again. Actually, the death of these little birds is probably due to starvation and not to cold acting directly; for Rowan[64] has shown that a small Canadian finch (*Junco hyemalis*) can, if provided with plenty of food, withstand blizzard temperatures ranging down to −52° F. As long as there is fuel, the body temperature can be kept up. It is probably the effect of frost upon the rest of the environment which is the serious factor in these cases.

6. The net result of periodic bad winters is therefore a periodic fluctuation in the numbers of many small birds. Many species of birds vary in numbers from causes which are largely unknown. Baxter and Rintoul[65] have studied for some years the fluctuations in numbers of breeding birds on the island of May in Scotland. They say : " Frequently it is possible to explain an increase in numbers of a nesting species by some change in environment, such as new plantations which afford convenient nesting places. But on the Isle of May no such obvious alterations in environment have taken place. Nevertheless the species come and go there, they increase and decrease, and the reasons are not by any means

always easy to discover. . . . We have written this paper to draw attention to the variations which take place in a limited area. It is a line of investigation which we think would well repay greater study, and which, if pursued in other areas showing different conditions, might yield sufficient data to make it possible to draw definite conclusions."

7. Irregular factors in the environment may sometimes act at long intervals but still have a tremendous effect on the numbers of animals. Wood-Jones [107c] says that the Cocos-Keeling Islands were visited by very severe cyclones in the years 1862, 1876, 1893, and 1902. These cyclones wrecked the settlement on the islands and caused much destruction to plants and animals. Similar destructive effects are produced by occasional drought years, which completely dry up ponds of less than a certain depth, causing partial or complete extinction of many species of aquatic animals in the ponds. Each drought is followed by a period of recovery, and it may take several years to reach normal numbers once more.

8. The irregular changes in the environment may be, on the other hand, favourable to the increase of animals, and so cause unusually large instead of unusually small numbers. For instance, the shipworm (*Teredo navalis*) was able to increase and spread in Holland in the years 1730-32, 1770, 1827, and 1858-59, owing to dry summers rendering the fresh-water regions more saline than usual.[66] It is sometimes the plant environment which changes suddenly. In tropical regions, *e.g.* India, there are certain species of bamboos which flower only at long intervals, and often simultaneously over large areas. This flowering gives rise to enormous masses of seed, and the unusual food-supply allows various species of rodents to increase abnormally, and sometimes to reach the dimensions of a serious plague.[69]

9. Periodic fluctuations in numbers may be studied most easily in mammals, since we possess rather accurate and extensive data about certain species whose fluctuations are extraordinarily regular in their rhythm. The best-known cases are among rodents, and of these the most striking are the Norwegian lemming (*Lemmus lemmus*) and the Canadian

varying hare (*Lepus americanus*). The lemming has for hundreds of years attracted attention by its periodic appearance in vast numbers. A certain Ziegler, who wrote a treatise which was published in Strasburg in the year 1532, was the first to make any record on the subject. He said that when he was in Rome in 1522 he heard from two bishops of Nideros a story about a small animal called the "leem" or "lemmer," which fell down from the sky in tremendous numbers during showers of rain, whose bite was poisonous, and which died in thousands when the grass sprouted in the spring. The story about the downfall from the skies was "confirmed" by Claussen in 1599, who brought forward new evidence of eye-witnesses who were "reliable men of great probity." It was not until the lemming had been described fairly accurately by Olaus Wormius, in 1653, and by later writers, that the real truth became known.[33] The Norwegian lemming lives normally on the mountains of Southern Norway and Sweden, and on the arctic tundras at sea-level farther north. Every few years it migrates down into the lowland in immense numbers. The lemmings march chiefly at night, and may traverse more than a hundred miles of country before reaching the sea, into which they plunge unhesitatingly, and continue to swim on until they die. Even then they float, so that their dead bodies form drifts on the seashore. This migration, a very remarkable performance for an animal the size of a small rat—MacClure called it "a diamond edition of the guinea-pig,"—is caused primarily by over-population in their mountain home, and the migrations are a symptom of the maximum in numbers which is always terminated by a severe epidemic; and this reduces the population to a very few individuals. After such a "lemming year" the mountains are almost empty of lemmings.

10. Owing to the striking nature of these lemming maxima we possess records of nearly every maximum for a number of years back, which enable us to find out the exact periodicity of the pulsations in lemming numbers.[23, 24] In recent years the maxima have occurred every four years, while in the middle of the nineteenth century they sometimes occurred at rather

shorter intervals of three or even two years. Now, by examining the records of skins obtained by the Hudson's Bay Company in Canada, it is possible to find out when the lemming years have occurred in Arctic Canada. This is made possible by the fact that the arctic fox numbers depend mainly upon those of lemmings, since the latter are the staple food of the fox. The curve of fox skins, which shows violent and regular fluctuations with a period of three or four years, can therefore be used as an index of the state of the lemming population in different years. When we compare the lemming years in

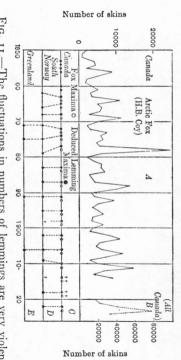

Number of skins

FIG. 11.—The fluctuations in numbers of lemmings are very violent and very regular, and synchronise in Scandinavia and Canada. Curve A shows the number of arctic fox skins taken annually by the Hudson's Bay Company in Canada. Curve B shows the number of skins taken each year in the whole of Canada, from 1920-1924. Diagram C shows the lemming years in Canada deduced from the fox curves A and B. Diagrams D and E show the known dates of maximum ("lemming years") of lemmings in Norway and Greenland (the latter being incomplete). (From Elton.[24])

Canada and Norway the curious fact emerges that they synchronise almost exactly in the two countries, and there is little doubt that the numbers are controlled by some common factor, which can only be climate. The other records which exist show that Greenland and the islands of the Canadian Arctic Archipelago also have lemming maxima at the same time as the others.

11. In the lemming we have therefore an animal which undergoes regular and violent fluctuations in numbers from year to year, which can be analysed into three main processes: first of all, the epidemic which kills them off when a certain

density of population is reached; secondly, the natural tendency to increase which leads to recovery of their numbers after the epidemic; and thirdly, the climatic factor which in some way controls this cycle. It might be thought that the lemming is exceptional among mammals in having such marked and regular fluctuations in its population. This is not the case. All small rodents, in all parts of the world for which there are any data, undergo periodic fluctuations of the order of three or four years. Wild rats, mice, hamsters, mouse-hares, gerbilles, etc., all appear to have this as their regular mechanism of number-regulation. The general fact has been noticed in a dozen different countries, *e.g.* England, Scandinavia, Central Europe, France, Italy, Palestine, Siberia, Canada, United States, South Africa, and to some extent Brazil and India. The larger rodents also undergo similar fluctuations in which epidemics play an important part. Squirrels in North America, Europe, and Asia have periodic maxima in numbers separated by intervals of five to ten years. Some of these maxima are associated with huge migrations like those of the lemming. In 1819 a vast army of grey squirrels swam across the Ohio River a hundred miles below Cincinnati,[101] while in 1897 a great swarm of the same sort passed through Tapilsk in the Ural Mountains : a solid army marched through for three days, only stopping at night, and they also swam across the river.[68] In Denmark [67] and Norway [33c] there are also periodic variations in squirrel numbers.

12. Rabbits and hares and jack-rabbits are subject to violent fluctuations in numbers in most parts of the world (*e.g.* North America from Alaska to Utah and California ; and Siberia). The best-known example is that of the varying hare or snowshoe rabbit (*Lepus americanus*), which inhabits Canada, and whose cyclical increase and decrease have long attracted attention owing to the effect that they have upon the numbers of valuable fur-bearing mammals which subsist on rabbits. The increase is partly due to the natural recovery after epidemics, and probably in part to climatic influences which speed up the rate of reproduction in certain years.[23]

Soper 26 gives a good description of the difference between the time of maximum numbers and of the minimum which follows the epidemic. He says : " It so happened, that upon my first visit to the West in 1912 the rabbit population was at its height. It was such a revelation after my eastern experiences, so startling, that the vividness of their abundance can never leave me. A certain brushy flat adjoining the White Mud River, south-west of Edmonton, yielded the initial surprise. It was grown to scrub willow, the common trembling aspen, and to

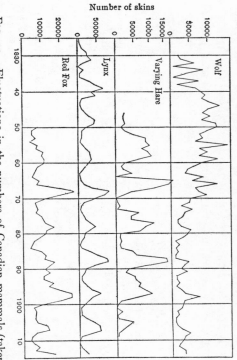

FIG. 12.—Fluctuations in the numbers of Canadian mammals (taken from Hewitt,8 who used the records of the Hudson's Bay Company). The figures for any particular year represent the effect of increase in the previous year, since the animals are trapped in winter. There is sometimes a lag of more than one year, owing to various trapping factors.

some extent with rank under-vegetation. The place was infested. I do not hesitate to say that over that tract of perhaps thirty acres hundreds of hares were found. October had come, without snow. The rabbits had already, wholly or in part, donned their snow-white livery of winter, and were consequently very conspicuous against the mellow brown of the autumn woods. At every turn during my ramble they popped up here and there and scurried for fresh cover. Not only in singles, which was astonishing enough, but often twos and even threes started up in wild alarm." This was the maximum abundance. Later on he says : " Eventually, evidence of the inevitable decline arrives. Empire among the rabbits as elsewhere has its rise and fall, and then is swept away. A strange peril stalks through the woods ; the year of death arrives. An odd rabbit drops off here and there, then twos and threes, then whole companies die, until the appalling destruction reduces the woods to desolation. . . . One year (1917) in the district of Sudbury, northern Ontario, the signs of rabbits were everywhere, but not a single rabbit could I start. It seemed incredible. Local inquiries disclosed that a little over a year before the *Lepus* population was beyond count. Now, as if by magic, they were gone. Needless to say, however, a few individuals survive the epidemic. These now, because of their paucity, are seldom encountered."

13. Marmots, muskrats, and numerous other rodents also fluctuate with a more or less regular periodicity. Seton [100] says " the muskrat's variation probably has relation chiefly to the amount of water, which, as is well known, is cyclic in the North-West."

In the case of lemmings, some mice, the varying hare, and the muskrat, it is practically certain that the cycle of numbers is partly under the control of regular cycles in climate ; and this is also probably the case with many other rodents about which we have at present little information beyond the fact that their numbers do fluctuate. At the same time there is no reason why such wave-motions in the population should not be automatic, in so far as each epidemic is followed by a period of recovery which would culminate in another epidemic, and so on indefinitely. It seems possible that this may apply to the rodents in some countries which show no regular or synchronous over-increase. The mouse-plagues in France are often local and always irregular, although sometimes they may occur simultaneously over large areas. Here variations in climate probably play at some times a more important part than at others. But on the whole, from what little we know about the matter at present, it appears that periodic fluctuations in the numbers of rodents are primarily caused by the irregular behaviour of their environment. After all, any species must either increase, remain constant, or decrease in numbers, and as there is always the danger of the numbers going *down* steadily if the balance is not struck exactly right, it seems reasonable to suppose that species would tend to be adapted to a steady increase in numbers, which is terminated at a certain point by disease or some other factor.

14. It is interesting to consider at this point an animal whose peculiar habits and scheme of social existence make the study of its numbers of extraordinary interest. This animal is the beaver (*Castor fiber*). The beaver has been hunted and trapped and studied so intensively for the last hundred years, that we possess a great deal of reliable information about its ecology. The statistics of the Hudson's Bay Company show that the numbers of beaver undergo no very marked short-period variations; although there was a general upward trend in number of skins during the early part of the nineteenth century, which was associated with the exploration and opening up of new parts of Canada.[84] This was followed by a considerable drop, consequent upon exhaustion of the natural stock of beavers through settlement and over-trapping. The reason for this comparative stability of the beaver population as a whole appears to be that it is almost entirely independent of short-period climatic variations. Since it uses the bark of trees for food the beaver is unaffected by annual variations of plant food-supply. It lives on capital and not on income— an almost ideal existence. Furthermore, although it is aquatic, the elaborate engineering feats which enable it to construct dams and houses (not to mention long canals for bringing food supplies from a distance) make it comparatively immune to the effects of annual variations in water-supply— unlike the muskrat, which is at the mercy of floods and droughts. Each colony of beavers lives in one place until the local supply of trees is exhausted, and when this takes place the animals apparently move on to some other locality. One would imagine that the beaver's habits, which cause it to live in isolated colonies, would make it difficult or impossible for any epidemic disease to spread throughout the population of a whole region (unless some alternative stage of a parasite were

carried by its enemies). This is actually the case. Epidemics are unknown among these animals, and, as a matter of fact, they have no very serious carnivorous enemies.[74] But local outbreaks of disease are probably an important check in numbers, since MacFarlane says,[78] "it is not an uncommon occurrence for hunters to find one or more beavers dead of disease in their houses or 'washes.'" In the beaver we find a rodent which apparently limits its numbers mainly by disease in the form of local outbreaks, and not by widespread epidemics; and which is unaffected by cycles of climate with a short period, so that the numbers of the population as a whole —apart from the effects of trapping by man—remain the same from year to year.

15. Fluctuations in the numbers of any one species inevitably cause changes in those of others associated with it, especially of its immediate enemies. This is well shown in rodent fluctuations. In Canada the arctic fox fluctuates with the lemmings, the red fox with mice and rabbits, the fisher with mice and rabbits and also fish, the lynx with rabbits, and so on. All these carnivorous mammals, which prey largely or partly on rodents, have periodic fluctuations of great regularity. This is true of weasels, ermine, martens, foxes, wolverene, skunks, mink, lynx, and others. The sable of Siberia depends to a great extent upon squirrels, and fluctuates accordingly. In all cases there is a lag in the increase of the carnivores owing to their larger size. Many birds of prey show similar fluctuations due to the variations in their food-supply, but here the question of numbers is complicated by migration, and it is rather hard to get at the real facts about changes in the whole population. But it can at any rate be said that the numbers of hawks and owls often vary regularly in any one place. One of the best descriptions of the enormous changes wrought in the abundance and habits of carnivorous birds and mammals by rodent cycles is that given by Cabot[119] at the end of his book on Labrador.

There is little doubt that insectivorous mammals and birds, such as shrews, moles, and warblers, are subject to considerable fluctuations dependent upon the abundance of their food, or upon other factors in their surroundings. Such variations have a regular periodicity in the case of the shrews [33a] at any rate.

16. Rodent fluctuations have always been recognised, although the universal existence of the phenomenon has not always been realised. It is less well known that the larger herbivorous mammals (ungulates) are also subject to periodic epidemics in nature, which are separated by much greater intervals of years than those of rodents, owing to the slower rate of increase of the former consequent upon their larger size. A rough idea of the rates of increase of the mammals of various sizes may be gained by the following figures. If one pair were allowed to increase unchecked, or at any rate were not subject to very severe checks (as by epidemics), a dense population would be produced by mice or lemmings in three or four years, by squirrels in five years, by hares or rabbits in about ten years, by sheep in about twenty years, by buffaloes in about thirty years, and by elephants in about fifty or more years. These figures are of course rough approximations, but they enable one to see why mammals of different sizes have different intervals between their epidemics and between their maxima.

17. Good evidence on the subject of fluctuations in wild ungulates is rather scattered and hard to obtain, since during the last few thousand years there have been many epidemics among man or his domestic animals which have become communicated to wild species and have thus interfered artificially with the natural balance of numbers of the latter. For instance, cholera and liver-rot seem to have been serious factors affecting the numbers of wild boars and deer on the continent of Europe during early times.[81d] But there are several bits of information which give us a glimpse of the way in which large ungulates used to regulate their numbers when living in an undisturbed state of nature. Brooks [80] says that on Vancouver Island the wild deer are subject to cycles of abundance and scarcity. These are followed closely by the cougar or mountain lion, which preys exclusively on deer, just as the lynx does upon rabbits. We do not know in this

PLATE I

Aerial photograph of a climax tropical forest in Burma (taken by Captain C. R. Robbins). Each lump in the photograph represents a forest tree some two or three hundred feet high. A ridge runs diagonally across the photo, and in the upper right-hand corner there are two white patches, which are landslides.

PLATE
II

The photograph (taken by Dr. K. S. Sandford in the Eastern Egyptian Desert in 1926) shows a typical limestone desert with blown sand and rock escarpments. This locality had had no rain for at least four years, and yet supported an open vegetation (probably owing to the supply of dew) and a fauna consisting of gazelles, ibex, together with a good many birds, lizards, insects, etc.

PLATE III

(a) A typical stretch of high arctic dry tundra, inhabited by reindeer, arctic fox, etc. (The photograph was taken in August, 1924, by Dr. K. S. Sandford, on Reindeer Peninsula, North Spitsbergen.)

(b) Drifting pack-ice near North-East Land (Spitsbergen Archipelago), with the bearded seal (*Erignathus barbatus*) lying on a floe. (Photographed by Mr. J. D. Brown, July, 1923.)

PLATE IV

(a) A semi-permanent pond in Oxfordshire (permanent except in drought years, when it dries up completely). Various successive zones of aquatic and marsh plants lead finally to grazed grassland on the right, and to elm trees on the left.

(b) Zonation of habitats on the sea-shore at low tide of Norway (Svolvaer), showing gradation from sea to land, through seaweeds (*Fucus*, etc.), drift-line, shingle, rocks, up to pasture or birchwoods.

PLATE V

(*a*) Zonation of plant communities on the Lancashire sand-dunes. The upper parts are covered by marram (*Psamma*), the lower parts by dwarf willow (*Salix repens*). Next comes marsh on the edge of the pond and finally aquatic plants, such as *Chara*, under the water. In the background can be seen the effects of secondary erosion by wind.

(*b*) Zones of vegetation on the edge of a tarn on a peat moor in the English Lake District. From left to right are (*a*) floating leaved plants (water-lily), (*b*) "reedswamp" of *Carex* mixed with floating pondweed (*Potamogeton*), (*c*) marsh with *Carex*, grading into sweet gale (*Myrica*) marsh, (*d*) bracken (*Pteris*), or coniferous woods.

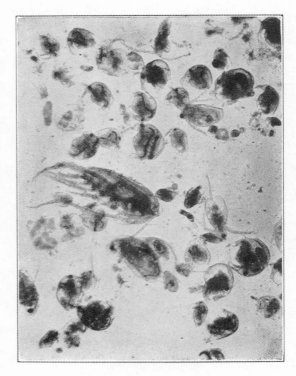

(*a*) A typical animal community in the plankton of a tarn in the English Lake District. Three important key-industry animals are shown: *Diaptomus*, *Daphnia*, and *Bosmina*.

(*b*) Effect of "rabbit pressure" on grass and furze on the Malvern Hills. The plants are closely nibbled by rabbits. The white web on the furze bush was constructed by a minute mite (*Erythræus regalis*, Koch. var.) which was present in enormous numbers.

PLATE VI

PLATE VII

(a) A guillemot cliff in Spitsbergen. The black streaks represent small flocks of guillemots flying away from their nesting places, disturbed by a gunshot. Each streak represents from ten to thirty birds. (Photographed by Dr. K. S. Sandford, 1924).

(b) Adelie penguin rookery on Macquarie Island. Amongst the penguins can be seen small areas of tussock grass and rock. (From Sir Douglas Mawson, *The Home of the Blizzard*, by permission.)

PLATE
VIII

A "cemetery" of walruses on Moffen Island, in latitude 81° N. These animals were slaughtered by early explorers for the sake of their tusks. In the foreground is an eider duck on its nest. (Photo by Dr. T. G. Longstaff 1921.)

case whether disease is the cause of this scarcity, but it seems very probable. Again, Fleming,[81e] speaking of the wild deer in England in 1834, refers to an epidemic " which from time immemorial had broken out at irregular intervals and swept off thousands of these animals." Percival[12k] says of the African zebra that " the free animal is carried off in considerable numbers by periodic outbreaks of disease, which has been traced to a lungworm."

18. It should be sufficiently clear by now that the numbers of many animals are subject to great fluctuations from year to year, and that, in the majority of cases which have been investigated, these fluctuations can be traced ultimately to pulsations or changes in the environmental conditions affecting the animals. If we follow up the implications of these facts it becomes possible to see why so many animals appear at intervals in vast plagues, often of great economic importance. Every herbivorous animal is adapted to increase at such a rate that enough extra individuals are produced to satisfy the requirements of its carnivorous enemies; while the latter, being usually larger, cannot increase so fast as their prey, and are accordingly adapted to a certain rate of increase which will not cause them to over-eat their food-supply. There are therefore small herbivores increasing fast, and larger carnivores increasing more slowly. The larger they are the more slowly they increase; this is a result of size-differences, but at the same time acts as an adaptation, in that it makes possible the existence of the food-cycle at all. Now, suppose one of these small herbivores—a mouse or an aphid—is suddenly able to accelerate its rate of increase, either as a population or as an individual. The change might be caused by a favourable winter which would enable the population to start in spring with a larger capital of numbers than usual. Again, it might be brought about by increased numbers of young being produced in each brood, as in the case of the varying hare or the short-eared owl. If the herbivore accelerates its rate of increase in this way, the carnivore, being larger and with slower powers of increase, is entirely unable to control the numbers of its prey any longer. And if the latter continues to increase at this

rate, its enemies will also do so, but will lag far behind and be too late to have any predominant effect on the numbers of the prey.

19. It seems, therefore, that the food relationships of animals result in the numbers being controlled in the majority of cases by carnivorous enemies, but that when disturbances in the environment cause a sudden acceleration in the rate of increase of some smaller species, its enemies no longer act as an efficient control. Actually the numbers of an animal are ultimately very often controlled by organisms smaller than itself, *i.e.* by parasites which produce epidemics. The following parallel may perhaps make this argument clearer. In ordinary weather the fire-brigades in a town are sufficiently numerous to keep down any outbreaks which may occur, but if there occurs a spell of very dry weather for several months there will be a sudden increase in the number and spreading powers of fires ; they will flare up more quickly and be more difficult to extinguish. In such an emergency the number of fire-brigades will be inadequate for dealing with the situation, and there will be a long delay before new firemen can be properly trained and equipped with helmets and hoses. Mean-while the fires will go on spreading and setting light to other houses, and in bad cases the whole town may be burnt down and have to be rebuilt again, as in the great fire of London in 1666. In just the same way, once a mouse population has " broken out " and escaped from the control of its enemies, it will give rise to a number of other mice, which in turn will increase and spread like the fire. The firemen (represented here by owls and foxes) cannot bring up and train more of their species in time to be able to stop the outbreak.

20. In the case of a bad fire the situation can usually be saved by a migration of fire-engines from neighbouring towns, but if it is a dry summer everywhere the fire-brigades will be busy putting out their own fires, and so unable to help any one else. Similarly, in a mouse plague there is often a huge migration of owls to the spot, which increase the chances of reducing the numbers. But here again the importance of climatic factors is seen, since the latter may cause plagues over

large areas, so that migration of enemies hardly takes place at all.

Although we have been taking mammals and birds as examples, since more is known about them and they are more familiar to most people, the principle outlined above applies universally in animal communities and is probably the explanation of most sudden plagues of animals. It results from the variable nature of the environment on the one hand, and the food-cycle mechanism of animal communities on the other.

Having shown the fact of fluctuations in numbers among many animals, and having discussed the causes of such fluctuations, we may next turn to a consideration of some of the effects which these fluctuations have upon animals in nature.

21. There are a number of curious and interesting consequences, of which we can only describe a few ; but it will become clear that variations in numbers play a very big part in the ecology of animals.

For instance, the food habits of many species depend not only upon the quality but on the quantity of their prey—a fact which is often lost sight of in experiments upon food preferences. As was pointed out earlier, the lower limit to the size of an animal's food is partly determined by the ease with which the latter can be caught ; and this in turn is tremendously affected by its abundance ; for if the prey is too scarce it takes too long to collect enough to satisfy the animal's needs. Time is a vital factor in the lives of most animals, and especially of carnivores. The animal must secure a certain amount of food every day or month or year, in order to survive and breed successfully. The seriousness of this factor varies in different species and at different times and places, but it is nearly always present during some part of the animal's life. It is said that the sheep in Tibet have to feed *at a run*, since the blades of grass are so scattered that only by being very active and energetic can the sheep get enough in a given time to support life.[144] With temperature-regulated animals the problem is of course more urgent, since they cannot usually allow the body-fuel to run too low, while with cold-blooded animals it is more a question of accumulating reserves of food big enough to carry through their breeding activities successfully. But in any case the problem remains.

22. The consequence of this is that as soon as a food animal sinks below a certain degree of abundance its enemies either starve or turn their attention to some other source of food. Fluctuations in numbers have therefore a potent influence upon the food-habits of animals. In fact, if several important key-industry species become suddenly very abundant or very scarce, the whole food-cycle may undergo considerable changes, if only temporarily. The various automatic balanced systems which exist will tend to bring the numbers, and therefore the food-habits, back in the long run to their original state. Another way of putting it is that the favourite food of an animal is not usually the most abundant one. Many animals have a definite scale of food preferences, depending upon quality, but if a favourite one becomes suddenly common the animal will abandon its previous food; and conversely, if a common food becomes suddenly scarce it may still continue to seek it up to a certain point, beyond which taste must give way to necessity. These variations in food-habits may lead to corresponding variations in the habitat to which the animal resorts. This is shown by the following example of the habits of the common rook (*Corvus frugilegus*), which is meant only to illustrate the idea, and not to prove anything final about the rook itself. During a particularly dull railway journey the writer noted down the number of pasture fields and of ploughed fields recently sown, and at the same time counted the number of fields of each kind which contained flocks of rooks feeding. The figures obtained were as shown in the table:

	No. of fields.	No. with rooks.	Per cent. with rooks.
Pasture	144	19	13
Sown ploughed ..	11	4	36

It will be seen that although there were comparatively few ploughed fields they appeared to be much more sought after than the pasture land (assuming for the sake of argument that the figures are sufficiently large to prove this—which they are not). But the absolute number of rook-flocks on pasture land was much greater than on ploughed land. If rooks had been much rarer birds we might have found that they were exclusively attached to ploughed fields at that time of the year.

23. Dugmore's description [45a] of the African buffalo (*Bos caffer*) in the Sudan shows in a very interesting way how the habits of an animal may depend upon its numbers. The wild buffalo which inhabits a large part of Africa used to be one of the most abundant large animals in that country, but in 1890 a frightful epidemic of rinderpest swept the continent, killing off, amongst other species, enormous numbers of buffaloes, and almost exterminating them in many parts of the country. Before the great epidemic the buffalo used to live in herds out in the open grassland and feed by day, but " for many years after [1890] the few remaining animals fed at night and retired to forests and dense swamps during the day." This was still the condition of affairs in 1910, but within recent years the buffalo have begun to increase again greatly and appear to have gone back to their former habits. This fact is confirmed by the observations of Percival [12c] and Chapman [85f] in Africa. It will be noticed that the African buffalo took about thirty-five years to recover its numbers. It is interesting, therefore, to find that the greater kudu (*Strepsiceros bea*) recovered in a much shorter time (ten to twenty years) from the rinderpest epidemic, although at the time it was almost wiped out. This quicker recovery is associated with the fact that it is a much rarer animal than the buffalo, and so requires less time to regain its maximum density of numbers.[12h]

CHAPTER X

DISPERSAL

1. When we are studying any particular animal or community of animals, we are brought up, sooner or later, against questions connected with dispersal : with the movements of animals in search of food, of shelter, or of their mates. This movement, on a large or small scale, is characteristic of animal communities, as compared with plant communities, and it forms a very important part of the lives of wild animals. Dispersal is such a large subject, and has so many sides to it, that

The study of animal dispersal involves (1) a large number of subjects besides biology, and we can only deal here with the general ecological aspects. (2) Most animal dispersal is directed towards the ordinary feeding, breeding, and other requirements of animals, not directly towards spreading the species, as shown (3) by the example of the capercaillie in Scotland. (4) The terms used in discussing the subject need to be carefully defined : the "spread," of any species involves "dispersal," "establishment of the individual," and "establishment of the species." These three phases sum up to control the "distribution" of a species at any one time. (5) Ecological succession plays a large part in slow dispersal of animals, but these do not necessarily extend the range of the species. (7) Besides active migration, many animals become dispersed by "accidental" means (which are often very definite and regular) ; they employ wind, water, logs, ships, seaweed, etc., and (8) other animals. Direct evidence on this subject is hard to obtain, and (9) is usually encountered by accident in the course of other ecological work. It is most important to publish such incomplete observations, as they may be unique. (10) Dispersal may take place as described above, through the ordinary activities of the animals, but where it is a definite large-scale phenomenon resulting in spread of the species, it may be either for the purpose of getting away from an overcrowded population or for getting towards some new place, or for both. (11) Much abortive colonisation is always taking place, partly owing to ordinary ecological factors being unsuitable and (12) partly to the difficulty of finding a mate upon arrival at a new place. (13, 14) Three methods of finding the right habitat are employed by animals : broadcasting huge numbers so that a few by chance reach the right habitat ; directive migration, by means of special instincts or tropisms ; and a combination of general broadcasting and local direction-finding.

146

it will only be possible here to give a general outline of some of the ways in which it affects animal communities. If we studied the factors which cause the onset of migration in animals like birds or lemmings, we should soon find it necessary to go quite deeply into psychology, in order to find out why animals come together into herds at all, why they sometimes become, as it were, hypnotised or obsessed with the migration impulse, and why they migrate in one direction rather than another. If we studied at all closely the reactions of insects, which lead them to seek a particular food-plant, or to lay their eggs in a particular place, we should soon become involved in remote branches of organic chemistry. Dispersal is especially attractive as a subject for study if only because it is so easy at any moment to escape from the purely biological side of the work, and seek a change in psychology, chemistry, meteorology, or oceanography. At the same time, we are chiefly concerned in this chapter with providing a general orientation towards the problem of dispersal, in so far as it forms a factor of the life of animals. It is a subject which requires clear thinking, since in the usual discussions about the present, and even more the past, distribution of animals, one notices a remarkable lack of comprehension of the extremely complicated processes involved in their spread, whether as individuals or species.

2. Since most plants cannot move about of their own accord, they usually have some special adaptation for the spreading of seeds or other reproductive products ; whereas animals, possessing the power of active movement on a large or small scale, do not have such highly developed special means of dispersal. Furthermore, their powers of movement are not usually employed for the direct purpose of spreading the species over the widest possible area. Although an animal like a rabbit, or an earthworm, or an earwig, has considerable powers of dispersal, these are directed towards its immediate needs of obtaining food, finding water, or a mate, or avoiding enemies, rather than towards the occupation of new territory for its own sake. The latter often takes place automatically as a result of the former type of activity, and this casual un-directed movement may become elaborated into regular

migrations; but these, again, are usually for the purpose of finding food, etc., and only result secondarily in the spread of the species. Of course this is not to say that some animals have not got very remarkable and specialised means of dispersal, which exist merely for the purpose of spreading the species. Such adaptations exist in nearly all sessile or very sedentary animals such as marine hydroids on the one hand, or spiders on the other. But the spread of species in an ordinary animal community (excluding coral reef and other marine littoral or intertidal sessile animal communities) usually consists of a rather vague and erratic shifting of the animals from one place to another, which after a number of years may sum itself up as a change in distribution. The cases of large-scale migration in a definite direction are either periodic (as in birds), i.e. they consist in a movement backwards and forwards between two or more places; or else are rather exceptional, as in the case of locust migrations or the outbursts of sand-grouse from Asia. The latter type of migration is very striking both from its size and also from its regular periodicity, but the number of species in which it takes place at all often is probably rather limited.

3. The idea with which we have to start is, therefore, that animal dispersal is on the whole a rather quiet, humdrum process, and that it is taking place all the time as a result of the normal life of the animals. A good example of this sort of dispersal is provided by the capercaillie (*Tetrao urogallus*), which became extinct in Scotland about 1770 but was reintroduced in 1837, and has spread gradually, so that it now occupies a very large part of the country again. The manner of its spread has been well described by Ritchie,[134] who gives a very interesting map showing the dates of its arrival at various places along the routes of its migration from each centre of introduction. The capercaillie is found almost exclusively in pine woods, and its migration was apparently caused by birds flying occasionally from one wood to the next, and so gradually occupying new territory. Parallel with this process went the increase in numbers, but it is almost certainly wrong to imagine that it was direct " pressure of numbers " that caused

it to migrate ; although, without accompanying increase in numbers, the species would not spread far. Ritchie says : "Dr. Harvie-Brown was of opinion that the capercaillie viewed prospective sites from its old establishments, and this very probable selection by sight, together with the fact that most of the woodland lay along the watercourses, would determine the capercaillie's dispersal along the valleys. Indeed, judging from the dates of the advent or establishment of birds in new areas, the valley systems ranked second only to the presence of fir woods in determining the course of the migrating capercaillies." The squirrel, which was also introduced into Scotland, behaved in a similar way, migrating along the valleys and frequenting the fir woods.[13e]

4. So far we have used the term "dispersal" in a rather loose way. It is necessary, however, before entering further upon the subject, to analyse the process of dispersal and define rather carefully certain terms which are usually used in a vague sense, or at any rate in different senses by different people. The following system of terms is used here, and whether or not it agrees with the definitions of other writers, it will at least be clear what is meant by them here. By "dispersal" is meant the actual migration or carriage of animals from one place to another : e.g. the floating of young spiders on streamers of gossamer, the flight of a migrating goose, the transport of water-mites on the legs of water-beetles, or the drifting along of jellyfish in the sea. When such an animal reaches its destination it may either die or survive. It usually dies, but if it does not, we say that it has "established itself as an individual." Thus, a large loggerhead turtle (Thalassochelys caretta) reached the coast of Scotland (Skye) in December, 1923, quite alive, and full of eggs.[125] This species is an inhabitant of tropical and subtropical seas. Again, in certain years large numbers of clouded yellow butterflies (Colias edusa) arrive in England from the Continent. But in such cases the individuals are unable to breed successfully, or else the young are unable to survive. The next stage is, therefore, that the animal must "establish itself as a species." It does this if it is

successful in breeding and starting a permanent population of its kind. Frequently a species may reach some new
place and breed, and may establish itself for a short time, but
then is wiped out. This often happens because the animal is
not adapted to some periodic factor which acts at fairly long
intervals, e.g. a very bad winter, or an epidemic. Or the
species may die out simply because its numbers and rate of
increase are not suitably adjusted to the new environment in
which it finds itself. For instance, after big invasions of
sand-grouse or of crossbills, pairs of the birds have been known
to breed in some localities for a year or two after their first
appearance.[114] But they usually die out in the end, and no
more are seen until the next invasion. Some botanists employ
the word " ecesis " (from the Greek word which means
dwelling at home) instead of " establishment." It is rather a
useful word, owing to its clear meaning and shortness, but at
the same time rather an ugly one.

The combined results of dispersal and of the establishment of the individual and then the species in a new place we
call the " spreading " of the species. The area covered by
it at any one time is its " distribution." These terms may be
summed up as follows :

Dynamic	Dispersal	
	Establishment as individual	} Spread.
	Establishment as species	
Static	Distribution.	

The loose and undefined use of words like dispersal, spread,
and distribution, by faunists and palaeontologists has led to
a good deal of confused thinking on the subject of past movements in the animal world, and it is well to be quite clear on the
subject, and to realise that the spreading of a species involves
at least three fairly clear-cut phases, and that successful spreading involves overcoming a series of obstacles at each of these
phases, so that it may be absurd to attribute the final result
to any simple factor or factors.

5. One of the important ways in which slow dispersal takes
place is through the migration or spreading of the environment, i.e. by ecological succession. The patch of vegetation

forming any particular habitat hardly ever remains for a long time in the same spot, owing to the process of ecological succession which is almost everywhere at work; and since succession takes place fairly slowly, any habitat carries along with it a complete set of animals, which thus gradually extend or change their range. This extension of range takes place slowly and quietly, and the species may end up by living in the same habitat but in an entirely different locality, as it were, " without knowing it." A simple example of this sort of thing is the southward extension of arctic species of animals during an ice-age in Europe. The animals must in many cases have changed their range with almost imperceptible slowness, as the physical conditions or vegetation gradually shifted southwards; often the process would be irregular, sometimes backwards and sometimes forwards, and over and over again patches of vegetation would be completely destroyed and wiped out. But on the whole the tundra zone, and the forest belt, would move south, and with it the fauna, until the latter had penetrated hundreds of miles south of its original home. The point is that such spreading would not necessarily be accompanied by any violent special migrations, except in so far as the latter were normally found for various other reasons.

Another example which illustrates the same thing on a smaller scale is the formation of " ox-bow " ponds in river valleys. On the flood plain of a place like the Thames Valley there are a number of ponds which owe their origin to the un-easy movements of the river in its bed. Arms, loops, or back-waters of the river become cut off as the latter shifts its course, and in the ponds so formed there are usually a number of molluscs or fish which would not have been able to reach the pond if the whole environment had not shifted. The examples of ice-ages and ox-bow ponds are only special cases of a universal and very important way in which species spread over the surface of the earth. The process consists of a combination of the migration of the environment and of the local move-ments of the animals in pursuit of their normal activities. Any new extension of a habitat, such as the edge of a pine wood, is immediately filled up in an almost imperceptible way.

When palaeontologists speak of the migration of animals into a new country, they usually have a vague idea of some species like the caribou walking from its original home in a large crowd, and arriving in force at the destination. What probably happens is rather the process described above, which is a great deal less spectacular and lasts over a very much longer time. Big-scale migrations do exist and of course attract attention from their very size, but it cannot be too much emphasised that the spreading of species is, in ninety-nine cases out of a hundred, a slow process intimately bound up with the local habits and habitats of the animals, and with ecological succession.

6. Among many animals the daily migrations in search of food and other necessities are more or less regular and well defined. Crozier [124] observed an amusing case of this sort among the molluscs of the coast of Bermuda. He found that *Chiton tuberculatus* (a species about nine centimetres in length) does not wander very far from its "roosting-place," but travels within about a radius of a metre from its home. One individual which was observed had a small *Fissurella* (length about 0.9 cm.) living on it, browsing upon the epiphytic growths which covered the valves of the *Chiton*. The *Fissurella* also wandered about all day over the *Chiton*, but always returned in the evening to the third valve!

Seasonal migrations of a regular nature occur among many animals besides birds. (For an up-to-date account of bird migration the reader may be referred to A. Landsborough Thomson's *Problems of Bird Migration*,[126] since the subject is far too large to be adequately dealt with here.) Instances which will immediately occur to the naturalist are the migrations of fish in search of their spawning grounds, or of herbivorous mammals in search of pasture. In Palestine a certain amount of excitement and discomfort was at one time caused by large numbers of migrating scorpions, which invaded the military camps that happened to have been planted in their path. In the same way it has been noted [127] that the common bandar or rhesus monkey and the Hanuman monkey of India trek in parties (formed of a mixture of the two species) from the plains to the hills of Nepal in the hot season, and return in the cold season, carrying their young with them.

Such seasonal migrations may be on any scale, ranging from the thousand-mile journeys of some birds, to the crested newt's spring migration to the nearest pond for the purpose of laying its eggs, or the even shorter journey made by aquatic worms and rotifers, when they repair to the hot surface-layers of conferva in ponds for the same purpose.

7. Amongst the various means of dispersal we can distinguish roughly between the more voluntary, or at any rate active, migration of the animal itself on the one hand, and the numerous means of " accidental " dispersal on the other. An enormous number of the smaller species of animals get about from one place to another by special means of transport other than their own legs, wings, or cilia. In certain cases, as when the larva has a special instinct or tropism which leads it to hang on to a particular animal, the dispersal may be apparently almost solely for the purpose of disseminating the species. Thus, the glochidia larvæ of the mussel have a special reaction to certain fish which causes them to shoot out the sticky threads by which they hang on to their temporary host. But at the same time, many of these cases involve the extraction of food from the host while the animal is hanging on, and so the dispersal is to that extent a secondary consideration.

Amongst the more curious means of dispersal is that adopted by the Emperor penguin, which breeds in mid-winter on the sea-ice on certain parts of Antarctica. Being a large bird, development takes rather a long time, and in spring the young birds are not always ready to take to the water by the time the sea-ice begins to break up and drift north. This difficulty was seen to be overcome by the birds taking their chicks to the edge of the sea-ice where it was beginning to break away, and waiting until the chunk upon which they were sitting freed itself and floated away northwards. The chicks were thus able to continue their development for some time longer before entering the water. In polar and temperate regions floating mats of seaweed carry large numbers of animals about attached to and living in them. Similarly, driftwood is an important agent of accidental dispersal. Beebe [128] states that on one log collected out at sea during the voyage of the *Arcturus*, he counted fifty-four species of marine animals, including numbers of crabs, fish, and worms. In modern times, ships act as the most gorgeous hollow floating logs and have been used by animals on a large scale.

8. Animals form the means of transport of a great many other species. Apart from parasites, permanent or temporary, there are a number of cases of accidental carriage which have been recorded. Thus, Kaufmann [129] noticed that the fresh-water ostracod *Cyclocypris lævis*, which he was keeping in an aquarium, had the habit of occasionally hanging on to the legs of certain water-beetles, so that it is possible that they may get about in this way from pond to pond. Water-mites have the similar habit of hanging on to all manner of insects (beetles, bugs, flies, etc.) in the larval state, and in this way getting carried to new ponds. There are a number of other examples which might be quoted, but to do so would obscure the fact that in an enormous majority of cases we have not the slightest idea in what way animals do commonly get about. Take the case of Polyzoa living in ponds. It is easy enough to conjecture that they may get about on birds' feet or some such manner, but very difficult indeed to find an actual case of their doing so. Only occasionally are we lucky enough to catch them in the act, as when Garbini [130] found the statoblast or winter egg of *Plumatella* on the beak of a heron. Often it is possible to obtain circumstantial evidence as to the methods of dispersal, evidence which usually takes the form of saying: "There are only two ways these animals could have got to this island (or pond, or wood), and this one seems the most likely." Thus Leege [131] studied the fresh-water crustacea in a pool of water on an island off the Friesian coast, an island which had only risen out of the sea during the previous ten years, and which must have obtained its fresh-water fauna from elsewhere by accidental dispersal. In 1907 there was no pool; in 1908 it was formed, and in 1909 von Leege found a large number of specimens of four species of common water-fleas (*Daphnia pulex, Simocephalus exspinosus, Pleuroxus*

aduncus, and *Chydorus sphæricus*). The island was frequented by enormous numbers of birds, and since these species of water-fleas are all littoral ones and furthermore form winter eggs which easily become entangled with birds' feathers, he concluded that the birds had been the transporting agent. This is usually the sort of evidence which one gets about accidental animal dispersal; the final proof is seldom forthcoming, owing to the difficulty of catching the animals in the act, especially when the dispersal may be a rather rare event in any case.

9. One is struck by the way in which the true and often quite important facts about animal dispersal, and especially accidental dispersal, are usually met with when one is working on something else in the field. In fact, it is almost impossible to spend any time profitably in a deliberate study of animal dispersal as such, except in certain well-defined directions, as in seasonal migrations of birds and fish, or parasites, or species whose larvæ regularly hang on to a definite host. Thus the writer has found a frog carrying a fresh-water bivalve (*Sphærium*) attached to its hind leg, while the real work on hand was the collecting of neotenous newts. And while studying the habits of rabbits in Herefordshire, he has suddenly noticed that hordes of young toads were migrating across the country in the direction of the local lake, and was surprised to find that they were making for a piece of water which they were unable to see, since it was hidden by a high wall. It seemed clear that they must be responding to the humidity of the air; but even that seems a little hard to believe, when one considers that they were more than a hundred yards from the water. It is on such facts as these that our knowledge of animal dispersal will always have to be built up, facts which are usually encountered by accident, in a casual way, and often in circumstances which make it impossible to follow up the problem any further. It is therefore especially desirable to publish such notes on dispersal, however fragmentary they may be, since there is not usually much chance of getting better ones for some years. And it may be noted here that there is a certain reluctance among zoologists to publish incomplete

observations, which is quite justifiable in the case of definite experiments, or of the descriptions of dead structures which will wait for you to observe them completely, and which can be checked by other observers if they are sufficiently interested. But in accumulating the life-history records of wild animals it is essential to publish any tiny fact about them which seems unlikely to be encountered in the near future, or which helps to provide another piece in the complete jig-saw puzzle which ecologists spend their time putting together. A good example in this direction has been set by C. B. Williams,[132] who has clearly demonstrated in his studies on the migration of the painted-lady butterfly the value of collecting together small and apparently isolated pieces of information, both from his own notes and from the observations of others, which, when carefully collated, throw a great deal of light on the process of dispersal.

10. We have said earlier that when an animal has accomplished the business of dispersal from its starting-point, the next process consists in the establishment of itself as an individual and then as a species, in its new home. There is, however, this exception to be noted. It seems highly probable, although difficult in the present state of our knowledge to prove conclusively, that many animals migrate on a large scale in order to *get away from* a particular place rather than to *go towards* anywhere in particular. That is to day, there are often very cogent reasons why a large section of the population should migrate somewhere else, the most common one being over-population. We see such pressure of numbers acting in the case of lemmings, locusts, sand-grouse, and aphids, when they suddenly depart in huge swarms for new lands. The point here is that the immediate reason for migration may be to relieve the situation in their normal habitat, and although it is all to the good for each species to establish itself from one of these migrating swarms in another locality (as happens in the case of the Caucasian locust migrations), yet the main effect of the migration is to remove a surplus population from the area usually inhabited by the species, and not to colonise other areas. It should, therefore, be borne in mind

that any large-scale migration of this sort may have two reasons, either to get away from the centre of distribution in order to prevent disaster through overcrowding, or to reach somewhere at the circumference and so extend the range of the species.

11. It is plain that an enormous wastage must occur while the establishment after dispersal is taking place, and that only a tiny fraction of the original emigrants will ever succeed in establishing itself even temporarily. We cannot do better than quote here the words of Wood-Jones,[107e] who had a peculiarly good opportunity of appreciating the factors in dispersal and establishment of species, in connection with the arrival of new animals and plants on the coral islands of Cocos-Keeling. He says : " Those creatures that are settled and established are the elect, and they are appointed out of a countless host of competitors, all of whom have had equal adventure but have gone under in the struggle, through no fault of their own. They are the actual colonists, the survivors of a vast army of immigrants, every one of which was a potential colonist." One particularly striking example was noted by him [107d] : occasionally huge flights of dragon-flies would arrive (belonging to the species *Pantula flavescens*, *Tramea rosenbergii*, and *Anax guttatus*), and would live for some time and feed ; but, owing to the absence of any permanent open fresh water on the islands, they never succeeded in establishing themselves permanently, although they actually laid eggs in temporary pools, which were not suitable for their breeding purposes. Another example of abortive colonisation on a huge scale was encountered by the sledging parties of the Oxford University Arctic Expedition whilst crossing the ice-cap of North-East Land in the summer of 1924.[22] One day in August, all three parties in different parts of the country encountered vast swarms of aphids (*Dilachnus piceæ*), normally found on the spruce (*Picea obovata*) of Northern Europe, together with large numbers of hover-flies of the species *Syrphus ribesii*. These insects had travelled on a strong gale of wind for a distance of over eight hundred miles, and had been blown in a broad belt across the island of North-East

Land (which is about the size of Wales). Since the surface is entirely covered with ice and snow, or else consists of very barren rocks with a high-arctic flora, there was not the remotest chance of either the aphids or the hover-flies surviving. As a matter of fact, the majority of them were wiped out by a blizzard which occurred three days later.

12. Abortive colonisation is happening everywhere in nature, on a smaller scale, and the two examples quoted above are given in order to drive home the fact that dispersal by itself may have absolutely no effect upon the distribution of a species, unless it is accompanied by effective establishment at the other end. The factors controlling the survival of the species have been dealt with in a general way earlier in this book, and it is therefore unnecessary to say more about them here. We may note, however, that one of the chief difficulties facing an animal upon its arrival is that of finding a mate, with which to co-operate in perpetuating its race. The chances of one individual copepod reaching a new piece of water by accidental dispersal may be small, and that of two individuals of opposite sexes doing so smaller still. But the chance of these two meeting and mating and bringing up young would often seem to be extremely remote indeed. We find, therefore, that animals which have to reach new places by definite migration or other special means of dispersal are either parthenogenetic during part of their life-cycle, or else migrate in large parties, like locusts. In the case of locusts there are well known to be special tropisms which cause the animals to stay together in swarms, and to follow the movements and flight of their neighbours, so that there is a good chance of a swarm of both sexes arriving at the other end.[38] In the case of most water-fleas the egg which is transported by accident hatches into a female, which is able by parthenogenesis to produce a huge population, amongst which males occur at a later stage. This is also a common method of colonisation used by parasites, whose life-history is often so risky that the chance of two animals of opposite sexes arriving together in one host is negligible.

13. With regard to the methods employed by animals to

find their proper habitat, when they are migrating or otherwise becoming dispersed, there are three important devices to be noticed. The first class of animals employs the method of broadcasting enormous numbers of young ones or even of eggs, with the result that some of them fall on stony ground and perish, while others reach the right place, where they have at least a very thin chance of establishing themselves. This method is employed by all those spiders which float away on gossamer threads when young; by locusts which start on migration with full air-sacs and loaded fat-bodies, and fly along until the air and the fat are used up, when they have to descend and can then only undertake small local movements [38]; and it is also used by an enormous number of sessile and sedentary marine animals, which produce free-floating larvæ that have only small powers of directive movement. This broadcasting involves a huge wastage of life, and is usually confined to young animals, except in cases which, like the locusts, are partly concerned with the relief of pressure in the home population.

The second method is to have some special reaction which enables the animals to find their suitable habitat; this is a very widespread method, and is much less wasteful than the broadcasting one. Pettersson [133] has studied the fluctuations in the Baltic herring fisheries and found that the herring probably only enter the Baltic when water of a certain salinity penetrates there. They follow the salt water, and refuse to go in water with a salinity of less than 32 or 34 per mille. At the present time the lower layers of the Baltic are hardly ever more than 28 per mille; but this condition is influenced by the tidal effects of the moon, so that it appears that in the past there have been at certain times invasions of the Baltic by comparatively salt water, such as still occurs in the Skagerak and Kattegat. His work on this problem makes it practically certain that the existence of a definite salinity-preference on the part of the herring, combined with peculiar tidal phenomena, has caused in the past regular fluctuations in the prosperity of the Baltic and neighbouring fisheries, whose period is about $18\frac{1}{2}$ years, with a superimposed longer period of 111 years.

14. On the other hand, some marine fish react to other factors than salinity. Thus, the mackerel in the Black Sea are said to migrate in response to a change in temperature. Whether this is the actual factor at work or not, it appears certain that they do not react to salinity.[134] With birds, the precise way in which they find their right habitat is not known, but in many cases they appear to know simply by a rather elaborate process of experience and memory, in others by some more mysterious sense of direction. Many insects have definite chemotropic reactions which lead them to choose the right habitat either for feeding or for egg-laying. For instance, Howlett[135] made *Sarcophaga* oviposit in a bottle with scatol in it, this being a decomposition product of albuminous substances ; while Richardson[136] made house-flies oviposit in response to ammonia together with butyric and valerianic acids. Barrows[137] found that the positive reaction of *Drosophila* to fermenting fruit was due largely to amyl, and especially ethyl, alcohols, acetic and lactic acids, and acetic ether.

But although a considerable amount of work has been done in this direction, and many more examples could be quoted, the fact remains that we are hardly able in any case to say how a particular insect does manage to find the right habitat to live in. That is to say, we can see that it inhabits certain vegetation and physical conditions, and that these are best suited to its physiological endowments, but it is hard to find out, and usually unknown, by what indication it is enabled to find these optimum conditions. This point has already been touched upon in the chapter on environmental factors (p. 40).

The third method of finding habitats when dispersal is in progress is by general broadcasting combined with local directive movement. This appears to be practised by some birds (homing pigeons amongst others) and by a number of insects. Thus, a butterfly may undertake a huge migration whose direction is only determined by the particular winds blowing at the time ; but if it arrives at any place at all like its original home it will then be able to find the right food-plant for its larvæ, by chemotropic or other means. It is possible

to perceive here the natural place of the study of animal behaviour and of tropisms in biology. Experimental studies in
this field are almost always conducted from the point of view
of physiology or psychology purely and simply; but there is
a large field for studying the natural significance of such
reactions with reference to the normal life of the animals.

There are several points about methods which (1) are of general importance in ecological work, e.g. (2) the recording of facts with an eye to the use to which they will be put in the future, and (3) the correct identification of species, which latter depends both (4, 5) upon a pleasant and comprehending attitude of systematists towards ecological work and (6) upon the collection of good systematic material by ecologists, who alone can provide the right data with the specimens. (7) The usual mistake among beginners is to under-estimate the number of animals of each kind. (8) Information from other people can be of great value if backed up by specimens of the animals concerned. (9) The carrying out of a biological survey involves various things : first, the listing of the main habitats, then (10) the collecting of the animals, together with careful habitat- and other notes, and finally (11) the construction of food-cycle diagrams, which (12) necessitates exhaustive study of the food habits of animals, a study which can be worked out in two ways, either separately or combined together. (14) The numbers of animals require special methods for their recording : one may use censuses in a given area or (15) in a given time, while (16) for recording variations in numbers it is advisable *not* to refer to "the usual " as a standard, but (17) to the numbers in the previous year or month, etc. (18) Finally, in publishing the results of ecological surveys it is desirable to include an index of species or genera, and (19) to employ certain special methods for recording the facts about food, etc.

1. ALTHOUGH the whole of this book is really concerned with methods of tackling ecological problems, rather than with an inexorable tabulation of all the important facts which are known about ecology, it is advantageous to draw together all the various lines of thought into one chapter, and to mention a few general ideas which may be of use as a background to ecological work. One of the most striking things about natural history facts is the haphazard way in which they are usually recorded. We are not referring so much to the fact that our knowledge of so many life-histories of animals has to be built up by piecing together fragmentary observations of different people, since it is impossible for any one person to be

lucky enough to work out a complete picture of the life and habits of any one animal in all its aspects and phases. The thing which strikes one is rather the way in which the observations are recorded, there being in many cases no principle followed. After all, it is impossible to describe an occurrence in the most useful way, without having some idea of how the information is going to be used. Adams [1a] has emphasised the importance of this in a very helpful chapter on ecological methods, and quotes Van Hise, who said : " I have heard a man say : ' I observe the facts as I find them, unprejudiced by any theory.' I regard this statement as not only condemning the work of the man, but the position as an impossible one. No man has ever stated more than a small part of the facts with reference to any area. The geologist must select the facts which he regards of sufficient note to record and describe. But such selection implies theories of their importance and significance. In a given case the problem is therefore reduced to selecting the facts for record, with a broad and deep comprehension of the principles involved, and a definite understanding of the rules of the game, an appreciation of what is probable and what is not probable."

2. The first point of importance is therefore to have a very clear idea of the use to which your observations will probably be put afterwards, whether by yourself or others. At the same time, of course, facts are constantly assuming an unforeseen importance in the light of later discoveries ; we are merely pointing out that it pays to try to look ahead and make records in such a way that they will be as intelligible and valuable as possible. These remarks may sound commonplace and superfluous, but an example will show the great importance of the point raised. When an ornithologist records the food of a particular species of bird, he very seldom troubles to find out the exact species of food-animals concerned. For instance, many food records merely refer to " mayflies " or " worms " or " *Helix*." Conversely, when an entomologist records the enemies of some caterpillar he will often enough refer to them as " small warblers," or if a worm-lover were to speak about the enemies of worms upon mud-flats he would probably

talk about "wading birds." Although it is possible to find out a great deal about the food-cycles in animal communities by working in terms of wider groups of animals than species, yet it is essential for a complete understanding of the problem to know the species of eater *and* eaten—a thing which we very seldom do know. If the ornithologist or entomologist took the small extra trouble of getting the foods accurately identified down to *species*, their observations would be increased about a hundredfold in value.

3. Although the number of observations about the food and enemies of various animals is prodigious, yet the majority of these data are just too vague to be of much value in making a co-ordinated scheme of the interrelations of animals. In other words, when one animal is seen eating another, it is very desirable to record the exact names of both parties to the transaction. The record of "green woodpecker eating flies" is of some use, as is the record of "Woodpecker eating *Borborus equinus*," but the ideal observation is "green woodpecker eating *Borborus equinus*." This point leads on to another important one, namely, the necessity for cultivating a proper "species sense." The extent to which progress in ecology depends upon accurate identification, and upon the existence of a sound systematic groundwork for all groups of animals, cannot be too much impressed upon the beginner in ecology. This is the essential basis of the whole thing; without it the ecologist is helpless, and the whole of his work may be rendered useless, or at any rate of far less use than it might otherwise have been, by errors such as including several species as one, or using the wrong names for animals. The result of such errors is endless misinterpretation of work, especially by people in other countries. It is possibly to this danger that we must attribute a certain lack of sympathy for ecological work, politely veiled or otherwise, which is sometimes met with among systematists. They realise that ecological observations are dependent upon correct nomenclature, and are therefore to some extent ephemeral, in cases where the latter is not yet finally settled. Added to this is the feeling that ecologists are rather parasitic in their habits and are to

some extent using other people (systematists) to do their work!

4. This feeling is natural enough, and arises from the fact that a systematist has two distinct functions : one is to describe, classify, and name all the species that exist (or have existed) ; the other is to identify specimens for other people, especially when elaborate technique and considerable skilled knowledge are required in the process. Now, it is only recently that the animal kingdom has begun to be completely explored. The systematist is still busy putting his own house in order, or, what is also often the case, putting in order the houses of other people who have died leaving them in a considerable mess. But in a great many groups of animals, we are really in sight of the time when there will be comparatively little purely descriptive work and classifying of species ; although, to the study of the exact limits of species and varieties, and what they are, there will never be an end. The point is that the system of classification is rapidly becoming standardised, and we shall, in the near future, be able at least to reach agreement (often arbitrary enough) as to what is meant by any specific name. It follows from what has been said, that the task of the systematist will become more and more that of the man who identifies specimens for other people, and less and less that of the describer of new species.

5. One of the biggest tasks confronting any one engaged upon ecological survey work is that of getting all the animals identified. Indeed, it is usually impossible to get all groups identified down to species, owing to lag in the systematic study of some of them (*e.g.* Planarian worms). The material collected may either be worked out by the ecologist himself or he may get the specimens identified by experts. The latter plan is the better of the two, since it is much more sensible to get animals identified properly by a man who knows them well, than to attain a fallacious sense of independence by working them out oneself—wrong. Also, in the majority of cases, there is simply not time for the ecologist to work out all the material himself, and it seems certain that nearly all primary survey work will in the future have to be carried on by

co-operation on a large scale between ecologists in the field and experts in museums. At the same time it is useful to know how to name the more obvious species of animals, and to know also where to find out general information about any particular group or species. A list of the works dealing with a number of groups of British animals is given in the bibliography at the end of this book. The list is necessarily incomplete, since in some cases (*e.g.* fresh-water planarian worms) no comprehensive work has been published ; while in many others, although the systematic work has been thoroughly done, the results are scattered in a number of periodicals, or are in relatively inaccessible foreign works, or else remain locked up inside the heads of experts who have not yet had time or opportunity to write the necessary monographs on their groups. Furthermore, in the list of works quoted, no complete treatment is attempted of protozoa, parasites, or marine animals.

6. The vital importance of good systematic work and the desirability of making it as far as possible available in a simplified form to working ecologists has been pointed out. We may now turn for a moment to the other side of this matter. It is very important that ecologists should, during the course of whatever work they are doing, pay attention to the collection of material which can be used for systematic studies. Ecologists often have unique opportunities for collecting large series of animals from one place at different times, and such series are often invaluable in helping to decide the limits of variation of different species. It is becoming more and more clearly realised that the habits and habitats of animals may form systematic characters quite as important as structural features, and that unless information of this type is accumulated in the form of good specimens with full data about habitats, etc., attached, there is no proof that one " species " does not contain a number of species, differing in such ways, but not in obvious structures. A striking example of this kind of thing is *Daphnia pulex*, whose life-cycle in Europe includes the formation of fertilised winter eggs which enable the species to tide over the winter until the following spring. In Spitsbergen there is also a *Daphnia* which is identical in structure

and habitat with the European form, except that it has the unusual power of forming winter eggs parthenogenetically, without the necessity of fertilisation by a male. There are in consequence no male *Daphnia* in Spitsbergen at all.[123] Another good example of the importance of field observations for distinguishing species is that of the British warblers, which are in some cases much more easily distinguished by their songs and nesting habits than by their appearance.

Now, the systematist is not usually a trained field naturalist, or, if he is, he lacks the knowledge of plant and animal associations which is required in order to define accurately the habitat of the specimens he is collecting. The ideal procedure would seem, therefore, to be that as full data as possible should be entered upon labels and handed over to the systematist with the specimens, and that a more detailed account of the environment, and in particular of the animal environment, should be published by the ecologist himself, who can employ, if necessary, some means of referring to the actual specimens collected.

7. We have dwelt at some length on the necessity of getting absolutely reliable identification wherever it is possible, because mistakes in this matter are one of the most fruitful causes of misunderstanding, while vagueness in description of an animal may render the most brilliant observations upon its ecology more or less valueless. The usual mistake made by beginners is in under-estimating the number of species in a genus and so becoming careless about checking all specimens obtained in order to get exact identification in all cases. Thus, suppose I record "*Daphnia pulex* eaten by sticklebacks"; there are two quite different species of sticklebacks, the ten-spined (*Gasterosteus pungitius*) and the three-spined (*G. aculeatus*), and since I had not distinguished them, you might begin to wonder whether I was aware of the existence of different species of *Daphnia* also. This element of uncertainty makes the value of the observation very small. In practice it often requires only a very small extra amount of trouble to collect a few specimens of the animal on which the observations have been made, or in the case of animals like birds and fishes, to look up in a book to see how many species there are and which

it was. It frequently happens that the person who chances to notice some fact of vital interest to the ecologist working on some problem is not the ecologist himself, but some other kind of biologist, or perhaps some one who is not a scientist at all. If the observations made by such people could be backed up by specimens of the animals, it would be possible to collect a vast amount of very valuable data.

8. It is worth bearing in mind that the ecologist can frequently get, in this way, facts which he would otherwise never come across at all by himself, and he should make every effort to enlist the help of other people to co-operate. In this connection it is worth while quoting from the rules which were made out by Dr. Levick for the use of non-biological members of the Northern Party of Scott's Antarctic Expedition.[121] They ran as follows : " Members are invited to write in this book notes on anything of interest seen by them relating to birds, seals, whales, etc., appending their initials and bearing in mind the following observations :

" (1) Never write down anything *as a fact* unless you are *absolutely certain*. If you are not quite sure, say ' I think I saw ' instead of ' I saw,' or ' I think it was ' instead of ' It was ' ; but make it clear whether you are a little doubtful or very doubtful.

" (2) In observing animals disturb them as little as possible. . . .

" (3) Notes on the most trivial incidents are often of great value, but only when written with scrupulous regard to accuracy."

These rules are useful for zoologists also.

9. Having given these general suggestions about ecological work, we will now consider the best methods of carrying out a general primary ecological survey of animal communities. Many valuable hints are contained in Tansley's book,[15] since to a large extent the methods of primary survey are essentially the same for plants and animals. The process of making such a survey is as follows :

First of all, have a general look round the area to be studied, and get an idea of the main habitats that exist, and in particular of the main plant associations. Don't bother about details yet, but simply try and get a grasp of the big habitats and habitat gradients. When you have made a list of the important habitats, come down more to details and subdivide these into smaller areas or zones, in the manner indicated in Chapter II. Thus, at this point your notes would be in the following form :

"The country can be divided roughly into the lower-lying parts which are cultivated, and the upper hilly parts which are not. The uncultivated area can be divided into three very distinct main zones :

" 1. Grassland.

" 2. Bracken, with scattered trees, forming a sort of bracken savannah.

" 3. Woodland.

" 1 and 2 are more or less abruptly separated, but 2 and 3 grade into one another at their margins owing to the complicated distribution of shrubs, such as bramble, and small trees, such as hawthorn."

Supposing we then took the woodland, the notes would go on something like this :

" There are several fairly distinct types of woodland.

" 1. Ash, with some sycamore.

" 2. Oak woods (which species ?), with hazel undergrowth.

" 3. Oak and sweet chestnut woods.

" These again vary much in undergrowth owing to the effects of fires, and felling, and age. N.B. This summer is dry enough to have caused grass fires, but the woods have not caught seriously."

If, then, we considered the oak-hazel wood, we might write :

" The oak wood can be divided into vertical strata :

" 1. Tree-tops—

" (a) Leaves.

" (b) Twigs and branches.

" (c) Under bark and rotten wood of branches.

" 2. Trunks—

" (a) Upper part with lichens (drier).

"(b) Lower part with mosses and liverworts (damper than last owing to run-off from the trunk, and height often only a foot, but varying according to the aspect).

"3. Hazel undergrowth, with some other shrubs.
"4. Herbaceous undergrowth.
"5. Litter of dead leaves, etc.; or 6. Moss carpet.
"7. Soil, underground."

10. This listing of habitats does not take very long to carry out, and is absolutely essential. Wherever possible the co-operation of a plant ecologist should be enlisted, in order that the plant associations may be accurately determined. But often it is sufficient to make lists of plants yourself from each of the habitat divisions (perhaps with the aid of some field botanist who knows the species well). Druce's *Botanist's Pocket-book* [99] is extremely useful for accurate identification of British plants. When this has been done, the next thing is to start collecting the animals from these different habitats, and in doing this there are several points to be borne in mind. First, it is vitally important to make as full notes as possible on the animals, and to record full details of the exact habitat, *e.g.* the species of plant on which they are found, whether they were on the upper or on the lower sides of the leaves, and any other observations made at the time, such as the reaction to light or rain, or food-habits, or numbers. The last is especially important. The data can either be written on a label with the specimen or, what is usually more convenient, a number can be placed in the tube or box containing the specimen and a corresponding number entered against the notes upon it in your notebook. A rather convenient method of making notes is to carry a few record cards such as are used for a card index, instead of the usual notebook. On the other hand, if a notebook is used, it is possible to take a carbon copy at the time, which may often be useful. Usually, however, it is impossible to make anything but very brief and rough notes in the field, and they have to be written up carefully at home afterwards. It is customary to warn students that they must make notes *on the spot*, and not afterwards. A trained

ecologist can, however, quite safely carry a lot of the details in his head and put them on paper at the end of the day. This is a matter of practice, and is a habit worth cultivating, as it saves much time and also makes it possible to do better work in wet weather, when note-taking is awkward. The chief time when it is best to take notes on the spot is when one is trying to prove something definite, since at such times it is very easy to forget the facts that do not agree with one's theories. It usually happens that a certain number of animals are found in odd places which do not fit in with any of the habitats originally listed, and these will necessitate some revision of the habitat-divisions you started with. Again, a number of animals are always occurring accidentally in the wrong habitat, and although they should be recorded carefully, the amount of detail as to their habitat need not be great. Discretion has to be used in this matter.

11. When a general idea of the distribution of the fauna has been gained, it is advisable to attempt the construction of a rough food-cycle diagram showing the relationships of the species. To do this accurately it is necessary to get the specimens identified, but a rough preliminary idea can be formed without knowing the exact species, although it is useless to publish such a diagram unless backed up by lists of the actual species concerned. It will be found necessary to organise a sort of ring of consulting systematists who are willing to work out material from the various groups of animals. It is a good plan, when sending large numbers of specimens (most of which will probably be quite common ones) to include, if possible, some which seem unusually interesting or rare, since in this way the expert who is working out the material for you will find it more interesting, and will be the more willing to help in the future. For details of methods of collecting and preserving animals the reader may be referred to the British Museum Handbook for Collectors,[112] which covers a number of animals, to Ward and Whipple,[97] who give excellent directions for most fresh-water animals, and to the various books given in the bibliography of special groups of animals.

It is *very important* that specimens should be killed and preserved in the appropriate way, as otherwise they may be useless for purposes of identification, or, at any rate, cause a lot of unnecessary trouble.

12. For constructing food-cycles there is only one method—the patient collecting of all kinds of information about the food and enemies of the species that are being studied. Direct observation in the particular place in which you are working is the best, since the food habits of animals are often very variable at different times and in different places. It may be convenient to summarise the various ways in which evidence about animals' food may be obtained.

(1) Watch the animal eating and, if necessary, take a specimen of its food (and of the animal itself). This is the type of evidence that is most difficult to obtain.

(2) Examine the contents of the crop or stomach or intestine. This may give good positive evidence but is useless for proving a negative, *e.g.* remains of butterflies disappear very quickly under the influence of the digestive juices of birds.

(3) Finding stores of food, etc.

(4) From a study of the animals and plants associated with it, deduce the animal's probable food. This enables the field of observation to be narrowed down. For instance, the writer saw a ptarmigan rise from a hillside, and on going to the spot where it had been, found that a number of seeds and flowers had been eaten from various plants. Since this bird was the only large herbivorous animal in the region, it was certain that the bird had eaten them itself.

(5) Experiments may be made to confirm such probabilities.

(6) Examine excretory products, *e.g.* castings of owls containing remains of small mammals, or droppings of terns containing limbs of crustacea.

(7) The structure of the animal will help to narrow down the field to a particular size of food.

(8) Note any food preferences, with reference both to quality and quantity.

(9) The amounts eaten per day are of great interest, *e.g.* counts of the number of animals brought to its nest by a bird in a given time.

(10) Finally, the numbers of two species will often give a clue to the fact that one is feeding on the other, *e.g.* birds attracted by an unusual abundance of caterpillars on oaks.

13. There are two ways of tackling the problem of food-cycles and community-organisation of animals. One way is to start with one particular species and radiate outwards along its various connections with other animals and follow the train of associations wherever it leads. This was the method described in Chapter V. It is a very fascinating form of ecological work, owing to the variety of interesting facts and ideas which are encountered, and it also has the advantage that it can be carried out without any very elaborate previous survey or listing of all the species of animal in the district. On the other hand, one may list all the animals and then subdivide them according to their place in the community—herbivores and carnivores, key-industries, terminal species, large and small, and so on. The separate food-chains can then be worked out, and in this way one gets a better perspective of the whole community. Perhaps a combination of the two methods would be the best procedure.

14. Another important subject about which something may be said here is that of numbers. The study of numbers is a very new subject, and perfect methods of recording the numbers and changes in the numbers of animals have yet to be evolved. In practice, we have to deal with two main aspects of this matter. The first question is as to the best way of taking censuses of the animal population at any one time, and the second is the question of recording changes in the numbers from one period to another. With regard to the first, a certain amount of work has been devoted to the methods of estimating the absolute numbers of various animals. Quantitative work on plankton has reached a very high degree of efficiency ; the usual method consists in doing counts of small samples from the whole collection and then multiplying by a factor to get the total

numbers present. The method of weighing material is also used. These methods are fully dealt with by Whipple,[96] and by Birge and Juday.[92]

One of the more important recent inventions, not described in these books, is the apparatus designed by Hardy,[103] which enables marine plankton to be collected continuously on a band of silk as the ship moves along at sea. This apparatus has already shown good results on the *Discovery* whaling expedition, and will provide information of great ecological interest. For by means of it, a belt-transect of the plankton can be made along any desired line, and variations in the fauna which are clearly shown can be used for correlation with the physical and chemical gradients in the sea, or with changes in the distribution of larger animals such as whales and fish.

Similar methods can be employed for soil-animals, and in fact for any animals which are sufficiently small and numerous to be susceptible to mechanical sampling and counting. The problem becomes much more difficult in the case of the higher animals like birds and mammals, which are more mobile, are constantly shifting their place of abode, and are, relatively speaking, so scarce as to make it impracticable to kill large samples and count them. However, it is comparatively easy to make accurate censuses of nests during the breeding season, and a good deal of work along these lines has been done in the United States. The reader may be referred to a recent book by Nicholson,[147] who gives an account of the methods of bird census successfully employed by him in England.

15. Grinnell and Storer[40] have successfully employed a different method of recording the numbers of birds. They say : " Instead of using a unit of area, we used a unit of time. Birds were listed, as to species and individuals, per hour of observation. In a general way this record involved area too. Our censuses were practically all made on foot, and the distance to the right or left at which the observer could see or hear birds did not differ, materially, in different regions. The rate of the observer's travel did, of course, vary some . . . also, in

some places, the greater density of the vegetational cover acted to limit the range of sight. But for each of these adverse features of the method there were certain compensations." One of the advantages of this method is that it gives a good idea of the relative numbers of animals in any association, and this is one of the most important types of fact about which we require information. It seems probable that the method will give information of great value, so long as a sufficient number of censuses are obtained by different people, in order to eliminate the effects of factors like the weather, time of day, rate of travel, etc.

16. In addition to censuses giving the average numbers of animals in different habitats, we require methods of recording the changes in numbers from month to month or year to year. Of course, a series of censuses of the kind described will provide this information, but in many cases there is not the time, staff, or opportunity for carrying out censuses of sufficient accuracy, in which the methods will remain the same as time goes on, so that the results are comparable. The chief difficulty of recording changes in the numbers of any animal which undergoes violent fluctuations in numbers is in finding a standard to which the abundance in different years can be referred. Such statements as " wasps are more abundant than usual " cannot be safely used, for two reasons. The first is that we do not know what " usual " means ; the second is that its meaning varies from year to year, and in the minds of different people. The latter is due to the fact that most people do not remember with any accuracy for more than about five years ; and also that more significance is attached to recent years than to earlier ones. The result of all this is that the word " usual " when applied to numbers may mean practically anything, according to the particular emotions, powers of observation, and strength of memory of the observer. The records of butterfly abundance in England given in the Phenological Reports of the Royal Meteorological Society, prove conclusively and surprisingly that butterflies are " scarcer than usual " in about one year out of every five ! It comes to this, that records referring to the " usual " are only of value

when they refer to years of very great scarcity or very great abundance. In intermediate years they are almost, if not quite, valueless.

17. The best method of recording the relative changes in numbers of fluctuating animals appears to be as follows : the numbers in any one year are referred to the abundance of the previous year. Thus we might say " small tortoiseshell butterflies more abundant this year than last year." If a continuous series of such records be made, we can then get a very clear idea of the *relative* abundance from year to year, and if there is any regular periodicity in the numbers, the maxima and minima will be quite easily distinguished. The advantage of this method is that it avoids the errors which arise when a fictitious average (" usual ") is used as a standard. Furthermore, most people can remember pretty clearly what the numbers were in the previous year, and so there is no danger of introducing a great error in this way. It is advisable to keep at least two separate records, one referring to the breeding season, and the other to the non-breeding season. Then the numbers in the breeding season of one year can be compared with those both of the breeding and non-breeding seasons of the previous years. The method is, of course, equally applicable to monthly or other variations in numbers ; its limitation is that it can only be used on fairly conspicuous animals. This method of recording changes in the numbers of animals requires if possible to be backed up by actual census figures in some years at least. In this way it would be possible to give the curve of fluctuating numbers an absolute value.

18. In conclusion, it is desirable to say something about publication of the results of ecological work and the best methods of presenting the facts so as to be of the greatest use to other people. We have already dealt with some of the more important errors into which it is possible to fall—insufficient description of the habitat and inadequate or inaccurate identification of species. There are one or two other points which are worth mentioning also. The first is that primary survey work and other ecological work dealing with large numbers of

animals belonging to different taxonomic groups has ultimately as its main use the elucidation of particular problems about individual species, by providing a picture of the biological surroundings of the animals. The result of this is that writers of ecological papers should aim at making their results as accessible as possible to the man who is working on one group or one species (*e.g.* some animal of economic importance). Now, it is usually impossible for such a man to pick out what he wants from amongst the great mass of facts contained in an ecological survey paper, with its huge lists of species. If, however, a short index to species or genera, or even families, is included at the end of the paper, it immediately increases its practical value to other biologists about a thousand-fold. Wherever possible, therefore, an index giving page references should be included, thus enabling the information about any one animal to be picked out with the greatest ease and saving of time.

19. Another method of presenting the results of ecological surveys, which has advantages, is that used by Richards,[18] and consists in tabulating the lists of species in the following way :

ANIMAL COMMUNITY OF TYPICAL CALLUNETUM.

Common name.		Latin name.	Food habits.	Habitat.
Moth	..	*Scythris variella* Stph.	Larva on *Calluna* and *Erica*	Adult hops about on bare ground.
Hover-fly	..	*Sphærophoria scripta* L.	Larva on aphides	Adult on flowers.

The best way of describing and recording food-cycles is another important problem to be faced in the publication of ecological work. Simple diagrams like those on pp. 58, 66, can be employed ; these are all right for showing general results, but when we wish to include a large number of species something more is required. Perhaps the most effective method would be to put in on the general diagram the group names, *e.g.* " aphids," together with a number referring to a list of the actual species in question—a list which would be too long to include on the diagram. It is also sometimes useful to include the relative sizes of the different animals, but here

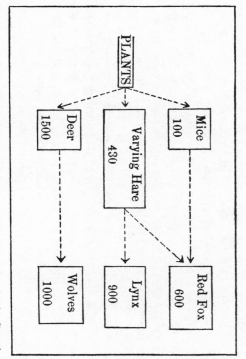

FIG. 13.—The diagram shows part of an animal community in Canada, and illustrates the method of including food-chains and the size of the animals in the same diagram. (The figures are lengths of the animals in millimetres—average for both sexes, tip of nose to base of tail.) This diagram should be compared with that in Fig. 12, giving the length of the period of fluctuation of the animals. (From Elton.[24])

one is usually hampered by lack of data or by the difficulty of finding a standard to which animals of different shapes can be referred.

CHAPTER XII

ECOLOGY AND EVOLUTION

(1, 2) Although the ordinary theory of natural selection appears, at first sight, to explain almost all the phenomena produced by evolution, the two greatest arguments in its favour being (3) the existence of so many perfect adaptations in animals and the difficulty of imagining how any but useful characters could spread in a population; yet (4) there are certain cases of colour dimorphism among animals which cannot be explained on the hypothesis of natural selection. Of these one of the most striking is the arctic fox, with its blue and white phases; (5) another example is the white-eared cob of the Sudan. In fact (6) it seems very likely that most so-called adaptive colours in mammals are not actually adaptive at all. (7) Furthermore, Richards and Robson have shown that it is highly probable that very closely allied species hardly ever differ in characters which are adaptive, although less closely allied species may do so. (8) These lines of evidence (from field observation on the ecology of the animals, and from systematics) make it very probable that there must be some process in nature which allows of the spread of non-adaptive characters in the population of a species. (9) The nature of this process will probably be revealed by ecological work on the numbers of wild animals, and (10) it is suggested that one factor in the spread of non-adaptive mutations is the expansion in numbers of a species after each periodic minimum in numbers, at which times the struggle for existence tends to cease or to become reduced. (11) Whatever may be any one's particular views on evolution, there is no doubt that ecological work is absolutely essential for a solution of certain aspects of the problem.

1. It may at first sight seem out of place to devote one chapter of a book on ecology to the subject of evolution. The reason for doing it is that the ecologist working in the field is continually being brought up with a sharp bump against the species problem. There are, in fact, certain aspects of the problem of the origin of species which can only be successfully tackled along ecological lines, and it is with these aspects that we shall deal in the present chapter, although it is here only possible to touch on some of the most important points. In order to make quite clear what part of the evolution problem is affected by ecological work, we must give a brief summary

179

of the present position of the subject. Every biologist accepts the fact that evolution has taken place. The problem which has not yet been really solved is the exact manner in which it has happened. The existence of vast numbers of undoubted and complicated adaptations in physiological, psychological, and structural characters makes it reasonably certain that Darwin's theory of natural selection must be essentially true, however we may disagree about certain parts of it. We start, therefore, by assuming that natural selection is a very important factor in encouraging the spread and perpetuation in the population of some of the genotypic variations which are constantly arising, and the cause of which is at present obscure. As we shall have occasion to point out that natural selection entirely fails to explain a number of phenomena in nature, it is well to be absolutely clear about the matter right at the start. The writer assumes that natural selection is an important factor in evolution, while at the same time holding that there are other agencies also at work, the nature of which will be best discovered by field ecological work on animals (just as Darwin and Wallace both discovered the existence of natural selection after an extensive experience of field work on animals). It should be further stated that the writer does not believe that there is as yet any conclusive evidence in favour of " the inheritance of acquired characters."

2. The ordinary hypothesis of evolution by natural selection may be summed up conveniently as follows :

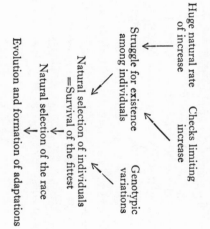

Huge natural rate of increase   Checks limiting increase

Struggle for existence among individuals

Natural selection of individuals = Survival of the fittest   Genotypic variations

Natural selection of the race

Evolution and formation of adaptations

Looked at from another angle, the process of formation of a new species can be divided up into three phases :

(1) Occurrence of a genotypic (heritable) variation.
(2) Spread of this variation in the population.
(3) Isolation of this new stock so as to form, ultimately, a new species.

The process of isolation does not necessarily come in in all cases, however, since a new variety might simply spread through the whole population of that species, and automatically change the whole stock.

3. With the origin of genotypic variations we are not here concerned. Ecology has, however, a definite contribution to make towards the study of the second and third phases. The usual Darwinian assumes that a variation which crops up singly, or at any rate rarely, has absolutely no chance of spreading in the population unless it is favoured by possessing some advantage over its fellows. This argument appears at first sight irrefutable. If a cod has a million eggs, and one of these eggs contains a new hereditary factor, what chance has this particular egg of growing up to the one of the two successful cods out of that million ? And if it did reach the breeding stage, and had a million young itself, only one of these, if any, would survive in the next generation. The Darwinian assumes that the deadly chances against any new variation spreading to any extent in the population can only be wiped out by the favourable influence of natural selection. If this is so, then all the characters possessed by animals—at any rate those which separate closely allied species—must either be of some direct use to the species (or to one sex in the species), or else they must owe their existence to the fact that they are intimately bound up in development with some other character which is useful, e.g. both might be products of the same hereditary factor in the egg. We see, then, that one of the great arguments in favour of the natural selection theory is the difficulty of any other hypothesis about the spread of variations, once they have arisen ; while another argument is that all animals are simply masses of adaptations.

4. So far, we have been arguing from one step to the next

until we are led by an apparently unassailable chain of reasoning to the existence of adaptations everywhere in nature. We must now leave arguments for a moment, and start at the other end by reviewing a few of the facts. In the front court of the British Museum of Natural History there are two cases which illustrate the beautiful colour adaptations of arctic animals to their surroundings, and will also serve to illustrate what we wish to point out here. There is in one case a group showing an arctic fox (*Vulpes lagopus*), some ptarmigan, and some ermine, in their summer dress of browns and greys, which match the surrounding vegetation with great exactness. In the other case the animals are shown in their winter dress of pure white, which makes them invisible against the snow. So far, so good. But further study of what is known about the field natural history of the arctic fox begins to reveal awkward facts, which do not fit in easily with this scheme of protective coloration, and in fact reveal a number of creaking joints in its harness. All over the arctic regions the arctic fox possesses two colour phases, one of which is brown in summer and white in winter, while the other is grey or black in summer, and "blue"— often quite black—in winter. The writer has seen a "blue" fox in summer which was the colour of a black cat, and startlingly visible against rocks and vegetation at a distance of a quarter of a mile. The blue and white phases occur equally in males and females, and interbreed freely, and in different parts of the arctic regions are found in various proportions in the population. In Iceland only the blue phase is found, while in Labrador it is rare. In Greenland, Alaska, and Spitsbergen both are common.[108] If the whiteness in winter is an adaptation, the blackness of the other phase cannot also be advantageous. If the black colour is adaptive, *how does the black survive?* We have in addition to reckon with the fact that in many parts of the arctic, the fox can have no possible use for its colour in winter, because it subsists at that season upon carrion left by bears, out on the frozen sea-ice, or if it is on land, it depends almost entirely on caches of animals collected and stored up in the autumn.

5. There are many similar cases of dimorphic forms which must have arisen by the spreading of colour varieties in the population, but which apparently cannot have been encouraged either by natural or sexual selection. Another case similar to that of the arctic fox, is the antelope called the white-eared cob (*Adenota leucotis*), which inhabits the steppe country of the upper Sudan.[85b] In this antelope there are two colour phases, one of which is light or tawny in colour and more or less matches its surroundings, while the other is dark or almost black. The interesting thing is that this colour dimorphism exists only in the male, the females being all light-coloured. The light individuals match their surroundings, the black do not. Taking the whole range of the species, there is an area in the middle of the range with black and light phases living together, and an outer zone with the light phase only. Examination of the horns showed that the differences in colour in the males were not due to age differences, as is often the case with such animals.

Here again, what at first sight seems to be an admirable adaptation in colour, turns out to be no better off than its companion phase which does not match its surroundings at all. Even if we assume that the colours are correlated with some other adaptation, the difficulty remains. There are, of course, a vast number of cases in which effective colour adaptation almost certainly exists (*e.g.* in the ptarmigan and in many insects), but in these cases there are never important dimorphic phases. But the fact that the adaptation exists in a number of cases does not in any way affect the fact that in certain other cases it *does not exist at all.*

6. It is rather interesting to find how emphatically nearly all naturalists who have had wide experience of wild mammals reject the idea of colour adaptation in these animals. Dugmore [45b] says : " The whole theory of protective coloration in the larger animals may be open to argument, but from my own observations in the field I am firmly convinced that practically speaking there is no such thing," while Roosevelt [86b] says : " In South America concealing coloration plays no more part in the lives of the adult deer, the tamandua, the

tapir, the peccary, the jaguar, and the puma, than it plays in Africa in the lives of such animals as the zebra, the sable antelope, the wildebeeste, the lion, and the hunting dog." Chapman [85e] gives evidence in favour of the same views.

It can always be argued about any of these animals that even if the colours are not directly adaptive they may be correlated in development with some character (perhaps physiological) which *is* adaptive. But such arguments cannot apply to species which are dimorphic, like the arctic fox or the white-eared cob. Similar colour dimorphism is found also in the Tibetan wolf,[43] the African lion,[108] and the American grey squirrel,[27] but is comparatively uncommon in mammals. In birds, however, it is often found. In the Galapagos Islands there is a hawk (*Buteo galapagoensis*) which has two phases (independent of age or sex), one of which is dark, while the other is pale buff,[36f] and a species of gannet (*Sula piscatrix websteri*) on the same islands which has two phases, brown and white.[36e] Many more examples could be given; a good deal of the very considerable evidence about birds has been summed up by Stresemann.[110] Exactly comparable colour dimorphism occurs in certain American dragonflies of the genus *Æshna*.[98]

7. There is another important line of evidence on the subject of adaptation which has recently been investigated very carefully by Richards and Robson and reviewed in a paper.[29] The gist of their conclusions is that very closely allied species practically never differ in characters which can by any stretch of the imagination be called adaptive. If natural selection exercises any important influence upon the divergence of species, we should expect to find that the characters separating species would in many cases be of obvious survival value. But the odd thing is that although the characters which distinguish genera or distantly allied species from one another are often obviously adaptive, those separating closely allied species are nearly always quite trivial and apparently meaningless. These two authors say, after reviewing the whole subject: "It thus seems that the direct utility of specific characters has rarely been proved and is at any rate unlikely to be common. Further-

more, since the correlation of structure, etc., with other characters shown to be useful does not at present rest on many well-proved examples, it cannot yet be assumed that most specific characters are indirectly useful. Thus the rôle of Natural Selection in the production of closely allied species, so far as it is known at present, seems to be limited. This statement is not to be taken as a wholesale denial of the power of Natural Selection. The latter is not in question when structural differences of a size likely to effect survival are involved. It is only the capacity of selection to use on a large scale the small differences between closely allied species that is unproved."

8. It seems probable that the process of evolution may take place along these lines : genotypic variations arise in one or a few individuals in the population of any species and spread by some means that is not natural selection ; this process, combined with various factors which lead to the isolation of different sections of the population from one another, results in the establishment of varieties and species which differ in comparatively trivial and unimportant characters. Later on, natural selection is ultimately effective, probably acting rather on populations than on individuals. Some such hypothesis seems absolutely necessary to account for the facts (driven home by ecological work in the field, and by careful systematic work at home) that, on the one hand, remarkable adaptations exist in all animals, while on the other hand the differences between closely allied species are not adaptive. This view is opposed to most of the current teaching about evolution, which tends either to exalt unduly or deny completely the power of natural selection, but it has the advantage of fitting the facts, which is after all not a bad recommendation. The most obvious question which arises is how a variation can spread in any population unless it is in some way favoured by natural selection. The process which, as we have pointed out, must happen and be happening, must be a mechanical one which allows of the spread of all characters indiscriminately. Any really harmful one would be wiped out soon enough by natural selection, and any really

useful one would be encouraged by natural selection. It is the indifferent characters with which we are concerned.

9. There is little doubt that it will be through ecological work upon the numbers of animals that this problem will be finally solved, and what we know already about the subject enables us to make certain suggestions. It has been shown in Chapter VIII, that nearly all animals fluctuate considerably in numbers, some of these fluctuations being very violent and often very regular in their periodicity. For our present purpose the important thing to bear in mind is the fact that at frequent intervals (frequent compared to the time which it would take for a species to change appreciably) the population of many animals is reduced to a very low ebb, and that this is followed by a more or less rapid expansion in numbers until the former state of abundance is reached once more. After a lemming year, with its inevitable epidemic killing off of all but a few of the animals, the arctic tundra is almost empty of lemmings. The same thing can be said of the snowshoe rabbit. One year the country is pullulating with rabbits, the following year you may hunt for a whole summer and only see one. There is usually a rather rapid expansion after this minimum of numbers. In a stream near Liverpool studied by the writer, the whole fauna over a stretch of three miles was wiped out during the summer of 1921, by a severe drought. Recolonisation took place from some deep ponds connected with the upper part of the stream, and after three or four years the population of molluscs, insects, crustacea, fish, etc., had regained its "normal" density. A similar destruction of the fauna took place in 1921 in a small branch of the Thames near Oxford, but by 1925 the animals had reached "normal" numbers again (through immigration and natural increase). The *Gammarus pulex* were very scarce in 1922, but had reached great abundance by 1925, when they were again practically wiped out, this time by an epidemic.

10. Now, if you turn back to the diagram on p. 180 you will notice that the argument contains a certain fallacy. The original theory says that all animals tend to increase, and at a very high rate, but are prevented from doing so by checks.

What has been said about fluctuations in numbers shows that such is not always the case. Many animals periodically undergo rapid increase with practically no checks at all. In fact, the struggle for existence sometimes tends to disappear almost entirely. During the expansion in numbers from a minimum, almost every animal survives, or at any rate a very high proportion of them do so, and an immeasurably larger number survives than when the population remains constant. If therefore a heritable variation were to occur in the small nucleus of animals left at a minimum of numbers, it would spread very quickly and automatically, so that a very large proportion of numbers of individuals would possess it when the species had regained its normal numbers. In this way it would be possible for non-adaptive (indifferent) characters to spread in the population, and we should have a partial explanation of the puzzling facts about closely allied species, and of the existence of so many apparently non-adaptive characters in animals.[23]

11. There are many objections to this hypothesis, which is chiefly mentioned here, not merely because it affords a possible solution of the problem of the origin of species, but because it illustrates the fact that ecological studies upon animal numbers from a dynamic standpoint are a necessary basis for evolution theories. Another important result of the periodic fluctuations which occur in the numbers of animals is that the nature and degree of severity of natural selection are periodic and constantly varying. For instance, at a minimum of numbers, rabbits will undergo selection for resistance to bad climate or ability of males to find females, while at times of maximum there will be different types of selection, *e.g.* for resistance to disease, and ability of males to secure females in competition with other males. All these points about adaptation, numbers, and selection, prove that ecological work has a very important contribution to make to the study of the evolution problem.

CONCLUSION

THERE is no getting away from the fact that good ecological work cannot be done in an atmosphere of cloistered calm, of smooth concentrated focussing upon clean, rounded, and elegant problems. Any ecological problem which is really worth working upon at all, is constantly leading the worker on to neighbouring subjects, and is constantly enlarging his view of the extent and variety of animal life, and of the numerous ways in which one problem in the field interacts with another.

In the course of field work one should have a rather uncomfortable feeling that one is not covering the whole ground, that the problem is too big to tackle single-handed, and that it would be worth while finding out whether So-and-so (a botanist) would not be able to co-operate with benefit to both, and that it might be worth while getting to know a little about geology or the movements of the moon or of a dog's tail, or the psychology of starlings, or any of those apparently specialised or remote subjects which are always turning out to be at the basis of ecological problems encountered in the field. There is hardly any doubt at all that this feeling of discomfort or conscience, or whatever you choose to call it, required in all scientific work, if anything more than routine results are to be produced, is most urgently required in ecology, which is a new science. Its methods require a wholesale overhauling, in order that the rich harvest of isolated facts that has been gathered during the last thousand years may be welded into working theories which will enable us to understand something about the general mechanism of animal life in nature, and in particular to obtain some insight into the means by which animal numbers are controlled. For it is failures in regulation of numbers of various animals which form by far the

188

biggest part of present-day economic problems in the field, and one of the aims of this book has been to indicate the lines along which the numbers of animals may be studied. The order of the chapters in the present book represents roughly the order in which it is necessary to study the ecology of animals. First, there must be a preliminary survey to find out the general distribution and composition of animal communities. Then attention is usually concentrated on some particular species, with the object of discovering what factors limit it in its range and numbers. As far as physical and chemical factors are concerned this can usually be done without reference to other species of animals, but one is practically always brought face to face also with biotic factors in the form of plants and animals. Ecological succession in plants, involving gradual migration of animal communities, has to be studied in this connection. When one starts to trace out the dependence of one animal upon another, one soon realises that it is necessary to study the whole community living in one habitat, since the interrelations of animals ramify so far. The study of an animal community is difficult, since so little work has been done on it; but there are certain principles which seem to apply to nearly all communities, and which enable the inter-relations of different species to be understood fairly clearly. It is only when the limiting factors, biotic and otherwise, have been appreciated that it is possible even to begin to study the numbers of animals and the ways in which they are regulated. One of the main facts which emerges from this study is that numbers do not usually remain constant for any length of time, but usually vary cyclically, sometimes with extraordinary regularity. Bound up with various aspects of animal com-munities is the question of dispersal of animals, which can be shown to be partly connected with the ordinary activities of animals, and partly with changes in numbers. Finally, what little we know about the regulation of numbers in animals, enables us to say that the problem of the origin of species can only be successfully solved by the aid of work on numbers.

The order of study of animals in nature therefore falls

naturally into a series of ecological phases : preliminary survey, factors (biotic and otherwise), animal communities, numbers, regulation of numbers, dispersal, and (if so inclined) origin of species.

Human ecology and animal ecology have developed in curious contrast to one another. Human ecology has been concerned almost entirely with biotic factors, with the effects of man upon man, disregarding often enough the other animals amongst which we live. Owing to the fact that most of the workers in this subject are themselves biotic factors, an undue prominence has been given in history and economics to these purely human influences. It is only recently, under the influence of men like Huntington [146] and Hill,[145] that the importance of physical and climatic factors in man's environment has become recognised. In animal ecology it has been entirely the other way about. Attention has been concentrated on the physical and chemical factors affecting animals, and if the biotic factors of vegetation and other animals have been studied, they have played quite a minor part in ecological work, or have been studied from the point of view of evolution, either to prove or to disprove the powers of natural selection in producing adaptations. As a matter of fact, we are now in a position to see that animals live lives which are socially in many ways comparable with the community-life of mankind, and if these resemblances be only considered as analogies, there yet remains the important fact that animal communities are very complicated and subject to regular rules, and that it is impossible to treat any one species as if it were an isolated unit, when we are studying its distribution and numbers.

I am ending this book with a diagram which attempts to illustrate in a rough way the relation of the various branches of ecology to each other, and to other branches of science. The diagram is based upon the order of the chapters in this book ; and it may serve as a reminder that ecology is quite a large subject.

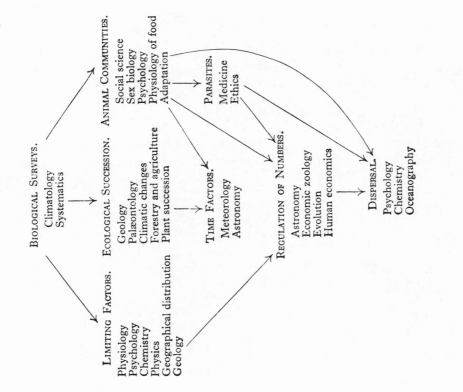

LIST OF REFERENCES

THE list of references given below is divided into three sections. The first comprises books and papers which will be found stimulating to the imagination and productive of ecological ideas. The list is necessarily rather arbitrary, and does not pretend to be anything more than suggestive. The second section contains references to various works on special subjects mentioned in the text, and to the examples which have been used to illustrate ideas in this book. Some of these occur also in the general works listed in Section I. The third section is devoted to works on the systematics and natural history of particular groups of animals, as explained on p. 166.

SECTION I.—GENERAL WORKS

1. ADAMS, C. C. (1913). Guide to the study of animal ecology. New York. 1a: p. 40. 1b: p. 82.

2. BUXTON, P. A. (1923). Animal life in deserts. London. 2a: p. 87. 2b: p. 115. 2c: p. 132.

3. CARPENTER, G. D. HALE (1920). A naturalist on Lake Victoria. London. 3a: p. 39. 3b: p. 53. 3c: p. 190.

4. —— (1913). Second report on the bionomics of *Glossina fuscipes* (*palpalis*) of Uganda. Reports of the Sleeping Sickness Commission of the Royal Society No. 14. 4a: p. 14. 4b: p. 37.

5. —— (1919). Third report on the bionomics of *Glossina palpalis* on Lake Victoria. Reports (as above) No. 17. 5a: p. 3.

6. CARR-SAUNDERS, A. M. (1922). The population problem. Oxford.

7. DARWIN, C. (1845). The voyage of the *Beagle*. London.

8. HEWITT, C. G. (1921). The conservation of the wild life of Canada. New York. 8a: p. 20. 8b: p. 210. 8c: p. 117. 8d: p. 232.

9. HOWARD, H. ELIOT (1920). Territory in bird life. London.

10. HUMBOLDT, A. VON (1850). Views of nature or contemplations on the sublime phenomena of creation. (Translated.) London. 10a: p. 199.

11. LONGSTAFF, T. G. (1926). Local changes in distribution. Ibis, London. 11a: p. 654. 11b: p. 656.

12. PERCIVAL, A. B. (1924). A game ranger's note-book. London. 12a: pp. 158, 160. 12b: p. 209. 12c: p. 254. 12d: p. 302. 12e: p. 332. 12f: p. 344. 12g: p. 345. 12h: p. 307. 12k: p. 303. 12l: p. 141.

13. RITCHIE, J. (1920). The influence of man on animal life in Scotland. Cambridge. 13a: p. 501. 13b: p. 508. 13c: p. 513. 13d: p. 272. 13e: p. 290.

14. SHELFORD, V. E. (1913). Animal communities in temperate America; Chicago.

15. TANSLEY, A. G. (1923). Practical plant ecology. London. **15a** : p. 97.

16. —— (1911). Types of British vegetation. Cambridge.

17. THOMSON, G. M. (1922). The naturalisation of animals and plants in New Zealand. Cambridge. **17a** : p. 27. **17b** : p. 81. **17c** : p. 88. **17d** : p. 154.

18. RICHARDS, O. W. (1926). Studies on the ecology of English heaths. Journ. Ecology, Vol. 14. **18a** : p. 246. **18b** : p. 264. **18c** : p. 249.

19. FARROW, E. P. (1925). Plant life on East Anglian heaths. Cambridge. **8a** : p. 38.

SECTION 2.—SPECIAL REFERENCES

20. SHELFORD, V. E. (1907). Preliminary note on the distribution of the tiger beetles (*Cicindela*) and its relation to plant succession. Biol. Bull. Vol. 14, P. 9.

21. Alaska, in the Encyclopædia Britannica, 11th edition. 1922.

22. ELTON, C. S. (1925). The dispersal of insects to Spitsbergen. Trans. Ent. Soc. London. August 7th, p. 289.

23. —— (1924). Periodic fluctuations in the numbers of animals : their causes and effects. British Journal of Experimental Biology, Vol. 2, p. 119.

24. —— (1925). Plague and the regulation of numbers in wild mammals. Journ. Hygiene, Vol. 24, p. 138.

25. SUMMERHAYES, V. S., and ELTON, C. S. (1923). Contributions to the ecology of Spitsbergen and Bear Island. Journ. Ecology, Vol. 11, p. 214. **25a** : p. 265. **25b** : p. 268.

26. SOPER, J. D. (1921). Notes on the snowshoe rabbit. Journ. Mammalogy, Vol. 2, pp. 102, 104.

27. —— (1923). The mammals of Wellington and Waterloo Counties, Ontario. Journ. Mammalogy, Vol. 4, p. 244.

28. SANDERS, N. J., and SHELFORD, V. E. (1922). A quantitative and seasonal study of a pine dune animal community. Ecology, Vol. 3, p. 306.

29. RICHARDS, O. W., and ROBSON, G. C. (1926). The species problem and evolution. Nature, March 6th and 13th.

30. DARWIN, C. (1874). On the structure and distribution of coral reefs. London. (2nd edition.) p. 20.

31. YAPP, R. H. (1922). The concept of habitat. Journ. Ecology, Vol. 10, p. 1.

32. LEVICK, G. M. (1914). Antarctic penguins. London. **32a** : p. 132. **32b** : p. 135.

33. COLLETT, R. (1911–12). Norges Pattedyr. Christiania. **33a** : p. 9. **33b** : p. 144. **33c** : p. 223.

34. WILKINS, G. H., quoted by WILD, F. (1923). Shackleton's last voyage. Appendix 2, p. 335.

35. MASSART, J. (1920 *circ.*). La biologie des inondations de l'Yser. Brussels.

36. BEEBE, W. (1924). Galapagos : world's end. **36a** : p. 72. **36b** : p. 92. **36c** : p. 122. **36d** : p. 222. **36e** : p. 321. **36f** : p. 287.

37. AUSTIN, E. E. (1926). The house-fly. British Museum (Natural History), Economic Series No. 1A. London. p. 41.

38. UVAROV, B. P. (1923). Quelques problèmes de la biologie des sauterelles. Ann. des Epiphyties, Vol. 9, p. 87.

39. CLEMENTS, F. E. (1916). Plant succession. Washington.
40. GRINNELL, J., and STORER, T. I. (1924). Animal life in the Yosemite. Berkeley, California. p. 22.
41. WILLIAMS, C. B. (1924). Bioclimatic observations in the Egyptian Desert in March, 1923. Ministry of Agriculture, Egypt: Technical and Scientific Service Bulletin No. 37. Cairo.
42. MACGREGOR, M. E. The influence of the hydrogen-ion concentration in the development of mosquito larvæ. Parasitology, Vol. 13, p. 348.
43. RAWLING, C. G. (1905). The great plateau. London. 43a: p. 316. 43b: p. 316.
44. RUSSELL, J. (1923). The micro-organisms of the soil. London. 44a: Chapter 9, by A. D. Imms. 44b: Chapter 5, by D. W. Cutler.
45. DUGMORE, A. R. (1924). The vast Sudan. London. 45a: p. 274. 45b: p. 281.
46. HADWEN, S., and PALMER, L. J. (1922). Reindeer in Alaska. U.S. Dept. of Agric. Bull. No. 1089.
47. HINTON, M. A. C. (1918). Rats and mice as enemies of mankind. British Museum (Natural History) Economic Series, No. 8, p. 45.
48. HOLLNSHED, quoted by MAXWELL, H. E. (1893). The plague of field voles in Scotland. The Zoologist, p. 121.
49. PIPER, S. E. (1908). Mouse plagues, their control and prevention. U.S. Yearbook of Agriculture, p. 301.
50. CRONWRIGHT-SCHREINER, S. C. (1925). The migratory springbucks of South Africa. London. p. 75.
51. HARMER, S. F. (1913). The polyzoa of waterworks. Proc. Zool. Soc., p. 426.
52. CHURCH, A. H. (1919). The plankton-phase and plankton-rate. Journal of Botany, June.
53. LORTET, L. (1883). Poissons et reptiles du lac Tibériade, etc. Arch. Mus. Hist. Naturelle de Lyon, Vol. 3, p. 106 (quoted in "Nature").
54. MARTINI, E. (1925). Neues über Wanderungen und Wirtswechsel bei parasitischen Würmen. Die Erde, Vol. 3, p. 24.
55. ROTHSCHILD, N. C. (1915). Synopsis of the British Siphonaptera. Entom. Monthly Mag., March.
56. CHRISTY, C. (1924). Big game and pigmies. London. p. 231.
57. WALLACE, A. R. (1913). The Malay Archipelago. London. p. 89.
58. WRIGHT, W. H. The black bear. London. p. 73.
59. STEWART, F. H. (1922). Parasitic worms in man. Nature, March 23rd.
60. BATY, R. R. DU. 15,000 miles in a ketch. London. p. 96.
61. PEARSON, T. G. (1924). Conservative conservation. The National Assoc. of Audubon Societies, Circular No. 9.
62. COWARD, T. A. (1920). The problem of the oak. Lancashire and Cheshire Naturalist, Vol. 13, p. 96.
63. ALLEE, W. C. (1923). Studies in marine ecology, No. 4. Ecology, Vol. 4, p. 341.
64. ROWAN, W. (1925). On the effect of extreme cold on birds. British Birds, Vol. 18, p. 296.
65. BAXTER, E. V., and RINTOUL, L. J. (1925). Fluctuations in breeding birds on the isle of May. Scottish Naturalist, Nov.–Dec., p. 175.
66. KOFOID, C. A. (1921). Report of the San Francisco Bay Marine Piling Survey; the biological phase. San Francisco.
67. HANSEN, H. J., and SCHIØDTE, I. C. (1878–1892). Zoologica Danica, Vol. 1. Pattedyr. Copenhagen. p. 88.
68. BREHM, A. E. (1897). From North Pole to Equator. London. p. 253.

69. Troupe, R. S. (1921). The silviculture of Indian trees. Oxford. Vol. 3, p. 982.

70. Powell, W. (1925). Rodents: description, habits, and methods of destruction. Union of South Africa, Dept. of Public Health. Bull. No. 321, p. 12.

71. Bulstrode, H. T. (1911). Report on human plague in East Suffolk. . . Rept. of Local Govt. Board on Public Health, etc. New Series, No. 52. Part I. London.

72. Niedeck, P. (1909). Cruises in the Behring Sea. London. p. 36.

73. Perkins, R. C. L., and Swezy, O. (1924). The introduction into Hawaii of insects that attack Lantana. Bull. Exptl. Stn. of Hawaiian Sugar Planters' Assoc. Entomological Series, Bull. 16.

74. Bailey, V. (1922). Beaver habits, beaver control, and possibilities in beaver farming. U.S. Dept. of Agric. Bull. 1078.

75. Elliott, H. W. (1884). Report on the seal islands of Alaska. Washington.

76. Saunders, J. T. (1924). The effect of the hydrogen-ion concentration on the behaviour and occurrence of *Spirostomum*. Proc. Cambridge Phil. Soc. Biol. Sci. Vol. 1, p. 189.

77. Vallentin, R., and Boyson, V. F. (1924). The Falkland Islands. Oxford. 77a: p. 336. 77b: p. 309.

78. MacFarlane, R. (1905). Notes on mammals collected and observed in the Northern Mackenzie River district. . . Proc. U.S. Nat. Mus., Vol. 28, p. 743.

79. Miall, L. C. (1897). Thirty years of teaching. London. p. 105.

80. Brooks, A. (1926). Past and present big game conditions in British Columbia and the predatory mammal question. Journ. Mammalogy, Vol. 7, p. 37.

81. Fleming, G. (1871). Animal plagues. London. 81a: Vol. 1, pp. 66, 96, 117, 119, 140, 228, 263. 81b: Vol. 2, p. 188. 81c: Vol. 2, p. 268. 81d: Vol. 1, p. 116; Vol. 2, p. 180. 81e: Vol. 2, p. 243.

82. Wilson, O. T. (1925). Some experimental observations of marine algal succession. Ecology, Vol. 6, p. 302.

83. Hofman, J. V. (1920). The establishment of a Douglas Fir Forest. Ecology, Vol. 1, p. 1.

84. Cooper, W. S. (1922). The ecological life-history of certain species of Ribes and its application to the control of the White Pine Blister Rust. Ecology, Vol. 3, p. 7.

85. Chapman, A. (1921). Savage Sudan. London. 85a: p. 42. 85b: p. 176. 85c: p. 284. 85d: p. 245. 85e: appendix. 85f: p. 190.

86. Roosevelt, T. (1914). Through the Brazilian wilderness. London. 86a: pp. 16–18. 86b: p. 94. 86c: p. 88.

87. Donaldson, H. H. (1924). The rat. Philadelphia. pp. 8, 9.

88. Schweitzer, A. (1923) On the edge of the primeval forest. London. p. 143.

89. Wenyon, C. M. (1926). Protozoology. London. Vol. 1. 89a: p. 385. 89b: p. 348 89c: p. 358.

90. Pearl, R., and Parker, S. L. (1923). American Journ. Hygiene, Vol. 3, p. 94.

91. Robertson, T. B. Experimental stuides on cellular multiplication: 1. The multiplication of isolated infusoria. Biochemical Journ. Vol. 25, p. 612.

92. Birge, E. A., and Juday, C. (1922). The inland lakes of Wisconsin. Wisconsin Geol. and Nat. Hist. Survey, Bull. No. 64. Scient. Ser. No. 13, p. 4.

93. MAWSON, D. (1915). The home of the blizzard. London. Vol. 2, p. 116.

94. HOWARD, L. O., and FISKE, W. F. (1911). The importation into the United States of the parasites of the gypsy moth and the brown-tail moth. U.S. Dept. of Agric. Bureau of Entom. Bull. No. 91.

95. DECOPPET, M. (1920). Le hanneton. Geneva.

96. WHIPPLE, G. C. The microscopy of drinking water.

97. WARD, H. B., and WHIPPLE, G. C. (1918). Fresh-water Biology. New York.

98. WALKER, E. M. (1912). The North American Dragonflies of the genus Æshna. Univ. of Toronto Studies, Biol. Ser. No. 11, p. 29.

99. DRUCE, G. C. (1922). The Botanist's pocket-book. London.

100. SETON, E. T. (1920). The arctic prairies. London. p. 109.

101. —— (1920). Migrations of the grey squirrel (Sciurus carolinensis). Journ. Mammalogy, Vol. 1, p. 53.

102. HARDY, A. C. (1924). The herring in relation to its animate environment, part 1. Ministry of Agriculture and Fisheries, Fishery Investigations, Series 2, Vol. 7, No. 3.

103. —— (1926). The Discovery Expedition. Appendix 2: A new method of plankton research. Nature, Oct. 16th, 1926.

104. BAKER, J. R. (1925). A coral reef in the New Hebrides. Proc. Zool. Soc., p. 1007.

105. VERSHAFFELT, E. (1910). Konink. Akad. von Wetenschappen te Amsterdam. Proc. Section of Sciences, Vol. 13, p. 536.

106. COUÉGNAS, J. (1920). L'aire de distribution géographique des écrevisses de la région de Sussac (Haute-Vienne) et ses rapports avec les données géologiques. Arch. Zool. Expér. Notes et Revue, Vol. 59, No. 3, p. 11.

107. WOOD-JONES, F. (1912). Corals and Atolls. London. 107a : p. 178. 107b : p. 241. 107c : p. 205. 107d : p. 307. 107e : p. 311.

108. LYDEKKER, R. (1903). Mostly mammals. London. p. 67.

109. HAVILAND, M. D. (1926). Forest, Steppe and Tundra. Cambridge.

110. STRESEMANN, E. (1925). Uber farbungs mutationen bei nichtdomestizierten vogeln. Verhandl. Deutsch. Zool. Ges. Vol. 30, p. 159.

111. HESSE, T. (1924). Tiergeographie auf ökologischer grundlage. Jena.

112. Handbook of Instructions for Collectors. Issued by the British Museum (Natural History). London. 1921 (4th edition).

113. BERG, B. (1925). Mit den Zugvögeln nach Afrika. Berlin.

114. WITHERBY, H. F. (1920). A practical handbook of British birds. London.

115. GERARD, C. (1871). Essai d'une faune historique des mammifères sauvages de l'Alsace. Colmar.

116. HUNTER, W. D. (1912). Journ. Econ. Entom. Vol. 5, p. 123.

117. CASSERBY, G. (1924). Where do wild elephants die? Journ. Mammalogy. Vol. 5, p. 113.

118. Phenological report of the Royal Meteorological Society for 1879. In Quart. Journ. R. Meteor. Soc., 1880.

119. CABOT, W. B. (1912). In Northern Labrador. London. Appendix.

120. SUMMERHAYES, V. S., and WILLIAMS, P. H. (1926). Studies on the ecology of English heaths. Part 2. Journ. Ecology, Vol. 16, p. 203.

121. PRIESTLEY, R. E. (1914). Antarctic adventure. London. p. 146.

122. THORPE, W. H. (1925). Some ecological aspects of British ornithology. British Birds, Vol. 19, p. 107.

123. OLOFSSON, O. (1918). Studien über die Süsswasserfauna Spitzbergens. Zool. Beiträge aus Upsala, Vol. 6, p. 422.

124. CROZIER, W. J. (1921). "Homing" behaviour in *Chiton*. Amer. Naturalist, Vol. 55, p. 276.

125. RITCHIE, J. (1924). The loggerhead turtle in Scotland. Scottish Naturalist, July-August, p. 99.

126. THOMSON, A. L. (1926). Problems of bird migration. London.

127. ANON. (1922). Trekking monkeys. The Field, February 25th, p. 248.

128. BEEBE, W. (1926). The Arcturus adventure. New York and London. p. 48.

129. KAUFMANN, A. (1900). Cypriden und Darwinuliden der Schweiz. Revue Suisse de Zool., Vol. 8, p. 209.

130. GARBINI, A. (1895). Diffusione passiva nella limnofauna. Memor. Accad. Verona, Vol. 71, Ser. 2, Fasc. 1, p. 21. (Cited in Archiv. für Naturgeschichte, Vol. 61, part 2, no. 3, p. 181.)

131. LEEGE, O. (1911). Die Entomostraken der Insel Memmert. 96th Jahresber. d. naturf. Ges. in Emden. (Cited by Wagler, E. (1923). Intern. Revue ges. Hydrobiol. und Hydrogr., Vol. 11, p. 45.)

132. WILLIAMS, C. B. (1925). The migrations of the painted lady butterfly. Nature, April 11th, p. 535.

133. PETTERSSON, O. (1912). The connection between hydrographical and meteorological phenomena. Quart. Journal Royal Meteorological Soc., Vol. 38, p. 173.

134. GALTSOFF, P. S. (1924). Seasonal migrations of mackerel in the Black Sea. Ecology, Vol. 5, p. 1.

135. HOWLETT, F. M. (1915). Bull. Entom. Research, Vol. 6, p. 297.

136. RICHARDSON, C. H. (1916). A chemotropic response of the house fly (*Musca domestica* L.). Science (New Series), Vol. 43, p. 613.

137. BARROWS, W. M. (1907). The reactions of the pomace fly (*Drosophila ampelophila* Loew.) to odorous substances. Journ. Exp. Zool., Vol. 4, p. 515.

138. TANSLEY, A. G., and ADAMSON, R. S. (1925). Studies of the vegetation of the English Chalk. Part 3. Journ. Ecology, Vol. 13, p. 213.

139. MANSON-BAHR, P. H. (1925). Manson's tropical diseases. London. pp. 508, 540.

140. KIRKMAN, F. B. (1910). The British bird book. London. Vol. 1.

141. FLATTELY, F. W., and WALTON, C. L. (1922). The biology of the seashore. London. p. 309.

142. MORLEY, B. (May, 1914). A larva plague in Deffer Wood, Yorks. The Naturalist, p. 151.

143. WHEELER, W. M. (1913). Ants. New York, Chapter 19.

144. LONGSTAFF, T. G. (1923). The assault on Mount Everest, p. 323.

145. HILL, L., and CAMPBELL, A. (1925). Health and Environment. London.

146. HUNTINGTON, E. (1925). The pulse of Asia. New York and London.

—— (1926). Civilization and climate. New Haven.

147. NICHOLSON, E. M. (1927). How birds live. London. Appendix.

148. SUMNER, F. B. (1925). Some biological problems of our southwestern deserts. Biology, Vol. 6, p. 361.

149. BELL, W. G. (1924). The great plague of London in 1665. London.

150. SCOTT, R. F., in Scott's last expedition. London, 1923, p. 74.

151. SCOTT, R. F. (1907). The voyage of the Discovery. London. Vol. 2, p. 224.

152. WILLIAMS, C. B. Memoir No. 1 of Dept. of Agriculture of Trinidad.

SECTION 3.—WORKS ON PARTICULAR GROUPS OF BRITISH ANIMALS

The system upon which the works listed below have been chosen is explained in Chapter XI. When it is stated that no suitable work exists on any particular group, this does not imply that no good systematic work has been published on that group, only that it is not to be found in English, or that it is inaccessible, or in a very scattered form, or else not sufficiently up to date to be very reliable. In almost all cases, except very recent publications, there has been published a good deal of additional work in various periodicals. The list is neither logical nor complete, but in spite of this, should be found useful to working ecologists. I am very much indebted to Dr. B. M. Hobby for revising the list of publications on insects.

Mammals

BARRETT-HAMILTON, G. E. H. and HINTON, M. A. C. (1910–1921). A history of British Mammals. London.

Birds

WITHERBY, H. F., and others (1920–1924). A practical handbook of British Birds. London. 2 vols.
HOWARD, H. E. (1907–1914). The British Warblers. London. 2 vols.

Reptiles

LEIGHTON, G. R. (1901). The life-history of British Serpents. Edinburgh.
—— (1903). The life-history of British Lizards. Edinburgh.

Fish

MAXWELL, H. (1920 *circ.*). British freshwater Fishes. London.

Molluscs

ELLIS, A. E. (1926). British Snails. Oxford.
'This book covers the gastropods of land and fresh water. There is no convenient and accurate work on the freshwater lamellibranchs, and the systematic position of a good many of the species is still in dispute.

Crustacea

WEBB, W. M. and SILLEM, C. (1906). The British Woodlice. London.
There is no comprehensive work upon the freshwater Crustacea of the British Isles, but much information can be obtained from Ward and Whipple's "Freshwater Biology," [37] which although dealing with American forms, is most useful, since many of the genera, and in the case of smaller forms, species, are the same in Europe and America. The following other works may be mentioned :
TATTERSALL, W. (1920). 'The occurrence of *Asellus meridianus* Rac. in Derbyshire. Lancs. and Cheshire Naturalist, Vol. 12, p. 273, (contains key to species of *Asellus*).
SARS, G. O. (1895). An account of the Crustacea of Norway, Christiania and Copenhagen. Vol. 1. (Includes an account of the genus *Gammarus*.)

GURNEY, R. (1931–33). British Fresh-water Copepoda. London. (Ray Society Publications.) 3 vols.

Watermites

SOAR, C. D., and WILLIAMSON, W. (1925–29). The British Hydracarina. London. (Ray Society Publications.) 3 vols.

Spiders

SAVORY, T. H. (1935). The Spiders and allied orders of the British Isles. London.
Does not give all the species. Includes some notes on mites.

Insects

IMMS, A. D. (1934). A general textbook of entomology. London. 3rd Edition.
An excellent work for general information and reference.

Thysanura.

WOMERSLEY, H. (1930). Contributions to a study of the British species of Machilidæ. Ann. Mag. Nat. Hist., Vol. 5, pp. 217, 278, 388.

Protura.

WOMERSLEY, H. (1927). Notes on the British species of Protura, with descriptions of new genera and species. Ent. Mon. Mag., Vol. 63, p. 140. See also *ibid.*, 1927, Vol. 63, p. 149 ; 1928, Vol. 64, p. 230 ; 1929, Vol. 65, p. 39.

Collembola.

WOMERSLEY, H. (1930). The Collembola of Ireland. Proc. Roy. Irish Acad., Vol. 39 (B), p. 160.

Orthoptera and *Dermaptera.*

LUCAS, W. J. (1920). A monograph of the British Orthoptera. London. (Ray Society Publications.)

Plecoptera.

MOSELY, M. E. (1932). A revision of the European species of the genus *Leuctra* (Plecoptera). Ann. Mag. Nat. Hist., Vol. 10, p. 1. See also papers by K. J. Morton in 1929, Ent. Mon. Mag., Vol. 65, p. 128 ; 1907, *ibid.*, Vol. 43, p. 107 ; 1902, *ibid.*, Vol. 38, p. 255, and 1896, Trans. Ent. Soc. London, p. 55 and 1894, *ibid.*, p. 557. A monograph on the British species of this order is much needed.

Psocoptera.

PEARMAN, J. V. (1926). A short account of British Psocids. Ann. Rept. Proc. Bristol Nat. Soc. for 1925, Vol. 6, p. 222.
PEARMAN, J. V. (1928). Biological observations on British Psocoptera. Ent. Mon. Mag., Vol. 64, pp. 209, 239, 263.
PEARMAN, J. V. (1932). Notes on the genus *Psocus*, with special reference to the British species. Ent. Mon. Mag., Vol. 68, p. 193.

Anoplura.

GRIMSHAW, P. H. (1917). The British lice (Anoplura) and their hosts. Scottish Nat., p. 13.

Ephemeroptera.

EATON, A. E. (1883). A revisional monograph of recent Ephemeridæ or Mayflies. Trans. Linn. Soc., Ser. 2. Zool., Vol. 3, p. 1.

Odonata.

COWLEY, J. (1935). The generic names of British insects, Part 3, p. 53 : The generic names of the British Odonata with a check list of the species. Roy. Ent. Soc. London.

LUCAS, W. J. (1930). The aquatic (naiad) stage of the British dragonflies (Paraneuroptera). London. (Ray Society Publications.)

LUCAS, W. J. (1900). British Dragonflies (Odonata). London. See also 1901, Entomologist, Vol. 34, p. 69, and 1904, *ibid.*, Vol. 37, p. 33.

Thysanoptera.

SPEYER, E. R. (1934). Some common species of the genus *Thrips* (Thysanoptera). Ann. Appl. Biol., Vol. 21, p. 120.

Hemiptera.

SAUNDERS, E. (1892). The Hemiptera-Heteroptera of the British Islands. London. Primarily a systematic work.

BUTLER, E. A. (1923). A biology of the British Hemiptera-Heteroptera. London.

This is biological rather than systematic, but brings up to date the systematic work which was done after the publication of Saunders' book.

JONES, H. P. (1928-30). An account of the Hemiptera-Heteroptera of Hampshire and the Isle of Wight ; with additional notes on British species not recorded for the county. (=A synopsis of the British fauna.) Ent. Rec. Supplement to Vols. 40-42, p. (1).

EDWARDS, J. (1896). The Hemiptera-Homoptera (Cicadina and Psyllina) of the British Islands. London.

LAL, K. B. (1934). The biology of Scottish Psyllidæ. Trans. R. Ent. Soc. London. Vol. 82, p. 363.

THEOBALD, F. V. (1926-29). The plant lice or Aphididæ of Great Britain. Ashford and London. 3 vols.

DAVIDSON, J. (1925). A list of British Aphides. London. Gives all references to the food plants of British species.

CHRYSTAL, R. N., and STORY, F. (1922). The Douglas Fir *Chermes* (*Chermes cooleyi*). Forestry Commission Bulletin No. 4. London. Gives keys to all British species of *Chermes.*

GREEN, E. E. (1927-28). A brief review of the indigenous Coccidæ of the British Islands, with emendations and additions. Ent. Rec. Supplement to Vols. 39 and 40, p. (1).

NEWSTEAD, R. (1901 and 1903). Monograph of the Coccidæ of the British Isles. London. (Ray Society Publications.) 2 vols.

Neuroptera.

KILLINGTON, F. J. (1929). A synopsis of British Neuroptera. Trans. Ent. Soc. Hampshire and S. Engl., No. 5, p. 1.

KILLINGTON, F. J. (1934). On the identity of *Hemerobius limbatellus* of British authors ; with a revised key to the British species of *Hemerobius* (Neur). Trans. Soc. British Ent., Vol. 1, p. 33.

KILLINGTON, F. J. (1934). On the life-histories of some British Hemerobiidæ (Neur). Trans. Soc. British Ent., Vol. 1, p. 119. A monograph of the British Neuroptera by the above author is shortly to be published by the Ray Society.

WITHYCOMBE, C. L. (1923). Notes on the biology of some British Neuroptera (Planipennia). Trans. Ent. Soc. London. 1922. p. 501.

Mecoptera.

HOBBY, B. M., and KILLINGTON, F. J. (1934). The feeding habits of British Mecoptera ; with a synopsis of the British species. Trans. Soc. British Ent., Vol. 1, p. 39.

Trichoptera.

McLACHLAN, R. (1874–1880). A monographic revision and synopsis of the Trichoptera of the European fauna. London.

Lepidoptera.

FROHAWK, F. W. (1934). The complete book of British butterflies. London.

MEYRICK, E. (1927). A revised handbook of British Lepidoptera. London.

SOUTH, R. (no date). The moths of the British Isles. London. 2 vols.

Coleoptera.

JOY, N. H. (1932). A practical handbook of British beetles. London. 2 vols.

FOWLER, W. W. (1887–1913). The Coleoptera of the British Islands. London. 6 vols.

DONISTHORPE, H. ST. J. (1931). An annotated list of the additions to the British Coleopterous Fauna made since the publication of the supplementary volume (VI) of Fowler's " Coleoptera of the British Islands." London.

BLAIR, K. G. (1934). Beetle larvæ. Proc. Trans. S. London Ent. Nat. Hist. Soc. for 1933–34. p. 89.

Strepsiptera.

PERKINS, R. C. L. (1918). Synopsis of British Strepsiptera of the genera *Stylops* and *Halictoxenus*. Ent. Mon. Mag., Vol. 54. p. 67.

Hymenoptera.

SAUNDERS, E. (1896). The Hymenoptera-Aculeata of the British Islands. London.

RICHARDS, O. W. (1935). Notes on the nomenclature of the Aculeate Hymenoptera, with special reference to British genera and species. Trans. Roy. Ent. Soc. London, Vol. 83, p. 143.

DONISTHORPE, H. ST. J. K. (1927). British ants. London.

DONISTHORPE, H. ST. J. K. (1927). The guests of British ants. London.

SLADEN, F. W. L. (1912). The humble-bee. London.

RICHARDS, O. W. (1927). The specific characters of the British humble-bees (Hymenoptera). Trans. Ent. Soc. London. Vol. 75, p. 233.

PERKINS, R. C. L. (1919). The British species of *Andrena* and *Nomada*. Trans. Ent. Soc. London, p. 218.

HAMM, A. H., and RICHARDS, O. W. (1926). The biology of the British Crabronidæ. Trans. Ent. Soc. London. Vol. 74, p. 297.

HAMM, A. H., and RICHARDS, O. W. (1930). The biology of the British fossorial wasps of the families Mellinidæ, Gorytidæ, Philanthidæ, Oxybelidæ, and Trypoxylidæ. Trans. Ent. Soc. London. Vol. 78, p. 95.

SWANTON, E. W. (1912). British Plant-Galls. London.
    This deals mostly with the Cynipidæ, but also includes galls formed by mites and flies.

MORICE, F. D. Help-notes towards the determination of the British Tenthredinidæ, etc. A series of papers on sawflies scattered through the Entomologist's Monthly Magazine, from 1903 onwards.

BENSON, R. B. (1935). The high mountain sawflies of Britain (Hymenoptera Symphyta). Trans. Roy. Ent. Soc. London. Vol. 83, p. 143.

MORLEY, C. (1903–1914). The Ichneumons of Great Britain. Plymouth, 5 vols.

For looking up biological facts about ichneumons : it is inadvisable to attempt identification of these insects, since it is a difficult and tricky process. For the same reason, no references are given in this list to the other parasitic Hymenoptera.

Diptera.

ANDREWS, H. W. (1931). British Dipterological literature. An annotated list of systematic monographs and books, published in English, dealing with British Diptera. Ent. Rec., Vol. 43, Suppl. p. (1).

In addition to the books and papers cited in this work the following papers will be found useful.

GOFFE, E. R. (1931). British Tabanidæ (Diptera). Trans. Ent. Soc. S. Engl., 1930, p. 43.

HOBBY, B. M. (1932). A key to the British species of Asilidæ (Diptera). Trans. Ent. Soc. S. Engl., Vol. 8, p. 45.

AUDCENT, H. (1932). British Tipulinæ (Diptera, Tipulidæ). Trans. Ent. Soc. S. Engl., Vol. 8, p. 1.

AUDCENT, H. (1934). British Liriopeidæ (Diptera, Nematocera). Trans. Soc. British Ent., Vol. 1, p. 103.

CURRAN, C. H. (1934). The families and genera of North American Diptera. New York.

GRIMSHAW, P. H. (1934). Introduction to the study of Diptera, with a key for the identification of families. Proc. Roy. Phys. Soc. Edinburgh, Vol. 22, p. 187.

Aphaniptera.

ROTHSCHILD, N. C. (1915). A synopsis of the British Siphonaptera. Ent. Mon. Mag., Vol. 51, p. 49.

THOMPSON, G. B. (1935). A revised list of the British Siphonaptera. Ent. Mon. Mag., Vol. 71, p. 181.

Leeches

HARDING, W. A. (1910). A revision of the British Leeches. Parasitology, Vol. 3, p. 130.

Flatworms

PERCIVAL, E. (1925). *Rhynchodemus britannicus,* n. sp., a new British Triclad. Quart. Journ. Micr. Science, Vol. 69 (new series), p. 343.

INDEX

INDEX

INDEX

INDEX